Efficient Electric Utility Operation

THE FAIRMONT PRESS, INC.

P.O. BOX 14227 • ATLANTA, GEORGIA 30324

EFFICIENT ELECTRIC UTILITY OPERATION

by O.C. Seevers, P.E.

Efficient Electric Utility Operation

Printed in the United States of America

ISBN 0-915586-59-2

Illustrations by Michael Lee

Second Printing: March 1984

CONTENTS

TO

F. J. Hardesty
R. H. McBeath
W. O. Percefull
J. F. Vest

Who are every bit as guilty as I

INTRODUCTION

This book is written for all electric utility engineers, managers, and superintendents. It describes methods which I have developed and used successfully in just about every aspect of utility operation outside the office. I even get into the hallway of the office.

I assume that you are already familiar with the usual dull and boring methods of operation which are standard in our beloved industry. So I concentrate on areas of operation for which I have devised and utilized unique methods which are not in use anywhere else in the world, to my knowledge.

To give you some idea of what you're in for, I'll give you a quick run-down of my topics.

I start by describing the embalmed state of our industry and how I have managed to stay awake.

I maintain that land rights are the first essential of our business and proceed to sound off on the subject.

Tree clearance is the bottomless pit into which we throw millions of dollars every year. Sure, we're going to pitch lots of dollars at this problem. I show you how to improve your aim.

The rain falls on the just and the unjust. Lightning falls almost entirely on transformer poles. I suggest you cooperate with it.

It is not too comforting to contemplate the fact that all of your tens of thousands of poles are quietly rotting. I suggest that you not cooperate.

Preventive maintenance, like contraception, is the only way to avoid unpleasant after-effects.

Inspection means "to look into." Be nosey. "Look into" certain elements of your system on a regular basis and record what you find.

System coordination would be easy if it weren't for people. I describe how to get from theory to practice.

Of course, switches should be properly maintained. But I use this as an excuse to tell an exciting tale.

OCRs and oil switches cost more than switches, so they should be maintained regularly, too.

Using contractors. This gives me a chance to sound off again on the sorry state of the utility industry and to suggest how to go about starting a revolution.

Trouble. We got trouble. Right here in River City. Or in whatever city you happen to operate in. I give you some ideas on how to meet trouble more than halfway.

I transform the storeroom into the glamorous service center at a cost you won't believe. I build a giant candy machine for storing poles.

Saving the best for last, I describe my unique efficiency rating system which I modestly call the best, simplest system in use anywhere in the universe.

I could go on and on. But I don't.

SYSTEM MAINTENANCE AND OPERATION

"Well, now, that's the way we've always done it."

How many sins have been committed in our utility industry with no more justification than that one stupid remark!

I have been amazed over the years to see huge sums of money—that's men's lives—wasted in following antiquated procedures in the operation and maintenance of transmission lines, distribution lines, substations and street lighting. And all that waste without giving the customer nearly as good service as he would get with less expensive, more effective methods.

Why, on God's green earth, do we do it?

We do it because we don't have the guts to change. No one wants to take the responsibility for instituting new methods which will produce absolutely no praise if they work and will for damn sure bring down waves of ridicule and scorn if they don't —or if they don't work as well as you claimed they would.

It's much safer to hide in the crowd, doing as nearly as possible the same as everyone else. Then if things go wrong you simply rely on the excuse that you were doing it just as it has always been done, the same as everyone else. If you don't stick your head up above the crowd you won't get it knocked off.

I find that pattern of thought repulsive and cowardly. I've spent my whole career doing just the opposite. True, I've had my head knocked off and my legs cut out from under me. I've suffered the slings and arrows of outrageous fortune and outraged management. But I haven't been *bored* to death.

I take the full credit for our success on the basis of one fundamental fact. That fact being my obstinate refusal to

3

compromise in the hiring of people. We've never been in position to outbid other companies and other industries for good engineers. But I didn't settle for second and third raters. I waited until I nailed top performers who decided to go to work for me for reasons other than pay.

I've always had a first-rate team eager to get things done and impatient with old, worn-out ideas. They've kept me on my toes. They are constantly full of ideas. *Ideas* are the basis of progress. Stomp on the first five ideas a young man comes up with and all you've got left is a bureaucrat. But give him some air and he'll turn things upside-down! He'll become a professional.

The world today is being run by amateurs.

A competent team of professionals is essential to produce and implement effective ideas. We have produced quite a lot of those ideas and I think they differ significantly from the run-of-the-mill methods to be noteworthy. I think they can be beneficial to you whether you swallow them whole or just nibble.

First, let's look at our purpose in all of this activity. Our purpose is to provide for our customers the best possible service within the limits of reason for the lowest possible cost.

We can't provide perfect service—that's impossible. Near perfect service is too expensive. It would cost more than the customer is willing to pay for it. Determining that point of diminishing returns is the science in which we specialize.

On one extreme is the new industry which blows into town with the demand that for their sensitive operation you must provide service with no blinks, no dips, no interruptions 24 hours a day, 365 days a year.

At the other extreme is the farm tenant house. They call up to report an outage a week after the lights go out.

That industry may as well be told from the outset that you don't provide no-dip, no-blink, no-outage service. On the other hand, you should be able to show them a record of service over several years without having to hang your head and scrape your foot as you do it.

How good should your service be? It ought to be as good, or better, than that provided in another comparable section of this country. The cost to the customer of that service should be less than it would be in another section of this country as a result of your efficient operation and maintenance.

What are the elements of operation and maintenance which we must deal with to provide service of this quality? I would list them this way:

- Land rights
- Tree clearance
- Effective grounding (Lightning)
- Pole maintenance
- Preventive maintenance
- Inspections
- System coordination
- OCR and oil switch maintenance
- Switch maintenance
- Using contractors
- Trouble
- Service Center
- Efficiency rating system

These items are listed at random, not in order of importance. Each produces a major effect on cost and quality of service.

LAND RIGHTS

Let's start with land rights.

If your approach to land rights is one of timidity and apologizing for inconveniencing the property owner, you are well on the way to a poor record of operation and maintenance. Adequate land rights are vital. Some soft-hearted (soft-headed) manager years ago agreed to a much-reduced easement width on a distribution line and promised that the owner's favorite tree would be given only limited trimming. How many outages have your customers been subjected to since then as a result? Another manager looks the other way while a house is built so that it extends onto the easement strip. How many more encroachments did you suffer as a result?

With each wife comes one mother-in-law. You may never get to like it, but you might as well accept the fact. If we are to provide quality electric service, we've got to have adequate land rights. If we establish standards early in the game, people get used to them. They may not like it, but they do learn to accept the fact.

By following these standards we can avoid appearing to be unfair to the property owners with whom we deal. Our local employees can monitor our easements without having to contend with a myriad of special cases.

It is not my purpose to suggest the wording which should be used in a proper easement. That's a job for an attorney. But it is important for the engineer to see to it that the attorney includes all the rights which are necessary for proper maintenance of the facilities involved.

I will make one suggestion on wording. "Agreement" is a much more palatable heading for a document spelling out land rights than "Easement." Would you rather enter into an agreement or give away easement rights?

Experience has proven the following easement widths to be necessary and reasonable.

- Cross-country transmission involving long spans — 150 feet
- Transmission in developing areas not involving long spans — 100 feet
- Distribution feeders through rural areas — 50 feet
- Distribution feeders in town or in subdivisions — 30 feet
- Single-phase lines which will never be three-phased — 20 feet
- Underground primary or secondary circuits — 20 feet
- Underground commercial service — 20 feet
- Underground residential service — 10 feet
- Underground low-voltage street-light circuits — 10 feet

All easements should include the:

- Right to construct, maintain, operate, replace, upgrade or rebuild pole lines or underground cable and appurtenances thereto.
- Right of ingress and egress.
- Right to trim and remove all trees on or adjacent to the easement strip necessary to maintain proper service.
- Right to keep easement strip free of any structure or obstacle which the company deems a hazard to the line.
- Right to prohibit excavation within 5 feet of any buried cable, or any change of grade which interferes with the cable.

- Right to install overhead or underground necessary wiring for street lighting that is requested and/or required, but no more than 5 feet from any lot line.

In the case of subdivisions, this information can be attached to the plat in the form of a "stick-on" before the plat goes to the planning and zoning board.

Now you have your easement rights signed, sealed and recorded. But unless your local people are trained and motivated to maintain vigilance over these rights they will do you little good. It is essential to explain to them, over and over, the importance of getting easements signed every time a line crosses private property. It is essential to explain to them, over and over, the importance of keeping an eagle eye for impending encroachments. It is not fair to allow a property owner to invest hundreds or thousands of dollars on a building which encroaches and then inform him it must be moved. We have to be on the lookout to catch such encroachments at the earliest date so that the property owner will be inconvenienced as little as possible.

An easement is a property right. There is no more justification for someone to construct a building on your company's easement than to do the same thing on his neighbor's property. The easement belongs to the stockholders of your company. It is not ours, as employees, to give away. We don't fight off encroachments for the sake of throwing our weight around. We should train our employees to avoid any such nuance. Rather, it should be explained that keeping the easement clear is essential to providing quality service to the property owner himself, and to his neighbors. Our employees must be sold on this concept. Too often they don't agree with our strict enforcement of easement restrictions so they are ineffective in their contacts with encroaching builders.

So why are we so snotty about encroachments? What does it hurt to let someone build a few feet over onto the easement?

A clear understanding of the answers to these questions should be implanted firmly in the mind of each employee who has any responsibility for protecting the company's easements. This includes servicemen, patrolmen, construction people, managers, engineers and customer service personnel.

We must maintain the integrity of our easements because not to do so subjects our facilities to several clear threats. First, buildings on the easement pose a fire hazard to our overhead lines. Fire and smoke boiling up through those lines can physically damage conductors, poles and equipment. It can cause the wires to move together resulting in short circuits. Fire and smoke can, themselves, shortcircuit the lines. Obviously, a building on fire directly under the wires will produce intense heat and damage to company facilities. The same building on fire, but located several feet away from the centerline of the electrical circuit, enhances the chance that the facilities will not be damaged, or at least not enough to interrupt service.

A building on the easement poses a threat to life. The lives of painters, roofers, antenna installers and, of course, the workmen constructing the building are jeopardized if the building is located close to overhead high-voltage lines. Once a shock and burn or fatality occurs, the company is subject to suit based in large part on its negligence in maintaining the integrity of its easement. If you allow a building on your easement so close that it violates code clearances, their attorney will really beat you to death.

When thinking about the dangers of a building on the easement we should picture the effect of conductor side swing. On a long span the conductor may be easily 50 feet off centerline during a strong wind. Even on short spans it is surprising to observe the extent to which conductors are often displaced. Take a distribution line on 8-foot crossarms. The side conductor starts out 4 feet off centerline. The wind can easily displace it another 5 feet. Your 30-foot easement is only 15 feet on each side of centerline and you've used up 9 feet, leaving only 6 feet between the conductor and the edge of the easement. Little enough!

You can't be sure when you are getting an easement signed

that a one-story building will be the only threat you will be faced with. Or that the one-story building in place at the time, or proposed, won't some day be replaced with a taller structure.

When an overhead conductor breaks it should fall on the ground, not on a building. It should fall on the ground so that fault current can flow back toward the source through the ground in sufficient quantity to operate protective devices—fuse, OCR, breaker. If it falls on a structure, the wood, brick, stone, or whatever material is used in the building will at least partially insulate the wire, reducing the amount of fault current which is permitted to flow. What current does flow will not likely be enough to operate protective devices, but will likely be enough to set fire to the building. There is also the real possibility that a conductor falling onto a building will be insulated, as described, and further, be left with an energized end hanging in reach of the public. This is a terrifying prospect.

A building on the easement impedes accessibility to your facilities. Besides being inconvenient, this can present a real hazard in performing necessary maintenance or repairs.

A totally obstruction-free right-of-way gives an appearance to the public which promotes respect for the dangers involved in engaging in any activity too close to the line. Every encroachment tends to obliterate that appearance and invites a lack of respect for the danger.

Any type of structure located on an easement where underground cables are buried obviously obstructs access to the cable.

The prohibition against digging within 5 feet of an underground cable is necessary to protect the cable from injury and also to protect the digger from coming into contact with energized conductors.

I am well aware that all I have been saying is so obvious to any reasonable person as to appear trite. However, experience has shown that these details need to be spelled out over and over again to the very people who shouldn't need to be constantly badgered with this information. Servicemen, managers, even engineers persist in taking the side of the property owner against the company whenever an encroachment arises. Everyone wants to be a "good guy" as opposed to the "bad guys" like you and me who carry final responsibility for maintaining the integrity of our right-of-way.

I chose to discuss land rights first from my list of the elements of maintenance and operation because you can't accomplish anything else on the list if you can't get to your lines to do it.

Trees and lightning are the two major villains we must constantly combat to maintain service. Their relative effects will vary in different sections of our country, but in my part of the country they run about neck and neck. Let's go after trees first.

TREES

I am not an expert on trees. I don't wish to be. We don't need any more experts on trees. We have enough. My knowledge of trees consists of the following information: Evergreens stay green, or sort of green, all the time. There are three kinds of evergreens. Cedars, pines, and others. Trees which don't stay green all the time are not evergreens. Included in this group are water maples, sugar maples, oaks, locusts, and others.

Some trees are strong and can be trusted to a limited extent. All other trees—don't turn your back on them.

The proper way to trim a tree that endangers an electrical circuit is with a chain saw at the ground line. If this is not feasible, move up a little.

Rural rights-of-way should be sprayed from a helicopter with whatever chemical your chemical company is pushing this year. Don't spray tobacco plants or small children, unless they are already running around claiming to be "environmentalists"— as opposed to brutes like you.

All tree work should be contracted to people stupid enough to do it. Such folk should not be allowed on your company's payroll. Everyone I hire for construction-maintenance work must be a potential first-class lineman or skilled laborer. No tree trimmer is. Therefore; all tree work should be contracted. Contractor tree men can obtain property owner permission. Company employees can't.

Company employees catch cold if they work in the rain, unless they are being paid at the overtime rate. Contractor tree men never get sick and never die.

Few contractor tree men can read. This keeps them from becoming a nuisance.

Company employees tend to fall out of trees. So do contractor tree men. But they don't go around complaining about it.

Contractor tree men are enthusiastic about their filthy work and take great pride in it.

I used to be a tree trimmer on a company crew. Most of us could read. The ones who couldn't were good trimmers. As a group, we were never able to compete with the contract crews. None of them could read.

In my opinion, there are two workers who are essential to an electric utility. The first-class lineman and the contractor tree trimmer.

Tree clearance should be done by contract because of the prejudices I have enumerated. Also, tree work is seasonal, requiring lots of manpower at one time and little or none at another. Company employees should not be subject to this uncertainty. Every time management decides to sell company stock, they cut off all tree trimming to artificially cut expenses, increase earnings and jump the price of the stock before the suckers who buy it catch on.

Hand clearing of new rights-of-way must be done when needed to clear a path for construction, not when convenient.

During the worst winter months manpower should be reduced to a minimum because trimming is more expensive during rotten weather.

When problems arise, you can call in the contractor superintendent and chew him out unmercifully. This is lots of fun. It keeps you from having a nervous breakdown from the frustration of knowing that when company forces screw up, you can only call in the superintendent and kiss his hand. You can maintain control over contract forces which is impossible with company forces.

So, let's use contract forces.

Now, what instructions do we give these contract forces?

I only give one instruction. "Don't let the trees burn or knock down my lines!"

"And do it cheap."

Money for tree clearance is limited. We could always use more. And use it to good advantage. We can't afford to waste any of it.

When I took control of our tree program 21 years ago, I found that 33 percent of our tree-trimming money was being spent on trimming and removing trees that were nowhere near our lines. "Public relations" trimming, it was called. I called it stealing. Managers were using our trimming crews to curry favor with influential local citizens. Not for any benefit to the company, although I'm sure they rationalized it that way. Actually, it was to enhance their own personal positions in the community. I suppose there was some ephemeral advantage to the company, but not in line with the cost.

I told the contractor and the managers that the next time I caught any tree work being done which was not directly connected with line clearance, I would have the crew's foreman fired. They didn't believe me. I caught it happening and had the foreman fired. That was the end of that. My tree-trimming budget went up 33 percent.

The purpose of utility tree trimming is to avoid outages caused by trees. We are not the Little Garden Club. We are not landscape architects. Our responsibility is to our customers, not to the ever-so-subjective opinions of every nut who thinks he, or she, is an expert on beautification. There is no way to trim a tree and leave it beautiful. All you can do is leave it trimmed. If a tree wants to be beautiful it shouldn't grow near a power line.

We shouldn't flim-flam the public by claiming that what we're going to do to that tree will improve its appearance. We should trim only what is necessary to insure that the line through that tree will not be damaged by that tree for two years. Then we'll come back and do it again. If two-year clearance runs into too much money, consider removing the tree altogether. Any tree that can be removed for three times the cost to trim, should be removed, if feasible.

How should the tree be trimmed? Everyone knows that any limb removed should be cut all the way back to the parent limb, or to a sizeable lateral branch. Nippers should never be used. All cuts should be made with a small power saw. After a tree has been properly trimmed once, only a few cuts will be required.

Only limbs which are threats to the lines, or will become threats during the next two years, should be removed. *Never NEVER* should a tree be "rounded over."

By leaving all limbs that don't threaten the lines, we are diverting the sap to them and away from the cuts, minimizing regrowth at the cuts.

Painting cuts is hogwash. Properly made cuts will heal themselves without merthiolate. On the first round of properly trimming your trees it is a good idea to paint the cuts to hide the fact that you've been there. Most people won't even know the tree has been trimmed if they don't see those "owl eyes" staring down at them.

If a tree is located directly under the wires, proper trimming will produce a "V" cut. If the tree is located to the side of the wires, only tree limbs are left whose growth can be directed away from the wires.

Sturdy, huge trees which tower over the wires require that all limbs in the vicinity of the wires which threaten the wires must be removed. But you can't afford to remove large, sturdy overhanging limbs which will not contact the wires unless they break. Let God trim it and you put the wires back up.

After you've done the best trimming job you can, consider replacing those 35-foot poles with 45s or 50s.

The importance of the circuit will determine how much you can afford to spend on tree clearance. On the other hand, after you've put a single-phase tap line to a tenant house back up for the third time in 12 months, you may decide it will be cheaper to do some whacking.

By now, you must have decided that I am a tree butcher, insensitive to the feelings of the public. If that were true, don't you think I would have been lynched by now?

Actually, our relationship with the public is excellent. We have few complaints. When we started the method of trimming I have described (stopped topping and rounding over) we had a flurry of complaints, some very heated. The first time around we sort of eased into the new method. The second time around we hit full force. After the first flurry, the complaints melted away.

The key to our success was this. I ordered the trimmer to trim "my way" or not at all. If we were refused permission we just moved on. But first, we asked the property owner to sign a form which simply stated that he was refusing us permission to trim. This made him think about the responsibility he was assuming. No one has ever signed two of these forms.

In Geneva, Switzerland, I saw trees trimmed in such a way that would have caused the Little Garden Club types in America to scream to high heaven. I saw the same thing in Paris, France. They believe in controlling the shapes of their trees to suit their ideas of what is pleasing to the eye. It is my personal opinion that a topped, rounded over, "shaped" tree is an abortion to the tree, to God, and to the public that has to view the ugly thing. Beauty is highly subjective. My opinion is worth no more than anyone else's opinion. No matter what method we choose to trim trees, it will offend someone. So why not choose the method that is the most effective in protecting our facilities, accepting the fact that it will surely offend a percentage of the population?

A perfectly symmetrical old oak in the middle of a pasture is a thing of utmost beauty. But so is a knarled old weather-beaten lightning-damaged oak that is anything but symmetrical. The point is, that my brand of trimming produces an assymetrical tree which has a certain beauty, at least to some people.

You can waste a lot of money trying in vain to please all tastes and fail to effectively protect your facilities. Better to aim at providing the best possible service to our customers and take a little flack until the public comes to accept this method of trimming.

Now let's look at how we pay the contractor and how we check to be sure our money is well-spent.

Cost-plus is the only feasible method of paying contract tree trimmers. Initial line clearing and helicopter spraying lend themselves to per-acre pricing and bidding. My interest here is not to discuss the financial aspects with which you are doubtless already familiar. My purpose is to look at method of payment as it affects results. So I will confine my comments to distribution tree trimming where this effect is likely to vary with alternative methods of payment.

What we are buying is not trees removed, or trees trimmed. What we are buying is trouble-free electric service. Trouble-free as far as trees are concerned.

This suggests that we might determine how much it is worth to maintain service free from tree problems for one two-year trimming cycle in a particular area. Find a contractor willing to contract to give you that performance for a price equal or less than what you have calculated it is worth to you. Now we don't care whether he cuts or trims the trees or not. He can hypnotize them, offer effective prayers, or inject them with a serum that will make them be good trees. He can't do anything that will cause us to get sued or that will give us a public relations gallstone. But the point is that we should limit our concern to providing ourselves with freedom from anxiety that trees are going to clobber our facilities.

Of course, we know that no one can provide the service I've described. Not total freedom from tree troubles. Maybe 90 percent freedom. Maybe 95 percent with a provision that tornado damage won't count against them. Maybe 95 percent for one price and 80 percent for half that price.

We could agree on a price and add a penalty clause for every customer-hour of outage suffered.

By now you can see that any such plan is unworkable. But the logic is valid. Before we dive into a plan of payments that is workable and before the inevitable seas of red tape close over our heads obscuring our vision, we need to visualize in our minds our true purpose. We should develop a plan of payment which will serve that purpose.

Let's go back to the beginning of my "unworkable" plan. I said we would determine how much we would be willing to pay for freedom from tree problems over a two-year cycle in a specified area. The area supplied by a local office would be a good unit to work with.

We would be willing to pay nothing, at one extreme. We would not be willing to pay a figure equal to the total revenue from that local office, at the other extreme. In the past we were willing to pay what we *did* pay for whatever trimming *was* done yielding *whatever* benefits we got. If that performance was satisfactory, surely the cost involved is a good indication of what we can afford to spend in the future. If not, we probably ought to plan to spring for more.

What about variables in cost? If you have been throwing money away trimming and removing trees which are no threat to your facilities and you have taken the pledge to quit "cold turkey," you can deduct that amount from your past expenses in the calculation of what your future expense should be.

Some years are heavy growth years. There are no average growth years and no light growth years. Each year produces a rate of growth ranging from above average to way above average. If minimum growth is 1, maximum growth 10, and average growth 5, *all* years will fall in the range of 7 to 10.

In my part of our country two-year trimming cycle is optimum. Whenever budget cuts force us to skip a year, we find that the cost of trimming the skipped area is doubled, if we maintain the clearance we would have had with the two-year cycle. Of course, we lost a lot of the trimming rights we had, because to drop the trees back where they were subjects us to charges of "butchery." Even so, it costs more and we get reduced line protection. In setting up a trimming budget with past trimming costs as basic data, we must adjust our figures for any section that has just been subjected to the budget-cutting wise guys.

I shall discuss trouble reporting and analysis elsewhere. For now, I shall assume that you have some means of measuring your tree-caused trouble in each operating district. It makes sense to shift tree trimming money in the direction of the district, or districts, with the worst records.

If you have been trimming trees with nippers, it will cost more per tree the first time around after you start doing it properly. Since you will probably not be allowed to spend a lot more than you have in the past, you will just get less trees trimmed. Four years from start, after the third trimming of a particular area, you will have all of your trees properly trimmed and you will be getting to every tree each time.

Using all of the information I have just listed, plus a heaping helping of common sense, you can determine what your initial budget should be.

The trimmers start working. You're paying them on a cost-plus basis. Do you just turn them loose and hope for the best? No. You must provide inspection and daily reporting.

Inspection is best provided by local company personnel—servicemen or local managers.

Daily reports list men, trucks, equipment, power saws, hours for each, and number of trees trimmed and number of trees removed.

The local inspector approves each daily report before forwarding it to the division tree-trimming boss. Two or three times a week he spends a half hour or an hour with the crew. He goes with the foreman sometimes and double checks the tree count. He verifies the number of men, equipment, etc., in use. He notes any out-of-service tools and equipment. He checks the quality of the work done. He makes sure trees which should be removed are removed, not trimmed. He coordinates the routing of the crew. In rare instances, he assists in getting permissions from property owners. He lets the foreman know that he is interested in his work, that he is checking his work and that he stands ready to help him. He makes his inspections at random times so the foreman never knows when to expect him. He reports what he finds to the division tree-trimming boss.

The division trimming boss, or one of his people, makes random field inspections to insure that the local inspections are done and that the information from those inspections is valid.

At the end of each quarter the daily reports are compiled into a quarterly report listing for each district and the entire division the money spent, number of trees trimmed, number cut, man-hours per tree cut or trimmed and cost per tree cut or trimmed.

This report is of limited value. The main value is in the comparision of man-hours per tree trimmed in one district against the other districts. It will flush out any crews who "go sour." The division average man-hours per tree trimmed should be less than .5.

The real measure of performance is whether a crew gets over its assigned area with quality trimming in the alloted time. That's what determines whether your customers receive service free from tree troubles, or not. And *that* is the purpose of all this excercise.

By compiling the time required to trim each area over a number of years we can develop a set of system maps with appropriate man-hours shown for each area. Taking into account the many variables previously described, we compare actual performance with that standard.

Any crew which doesn't measure up is replaced.

Once a year, the contractor has all the men in for a two-day school. One "professor" is our division trimming boss. On the final evening they get a banquet, preceded by a happy hour. After dinner they get, and make, speeches about the year's performance. I make a speech in which I try to compliment them and inspire them. I stress the importance of the work they do and the particular importance of their part in it. I wish our company employees demonstrated half the enthusiasm these men do. What pride they take in their unglamorous work!

This pride is the key to a successful tree program. The Mickey Mouse daily reports don't mean a lot to us so far as providing essential information, but they do help sustain this pride because, along with the inspections, they continually remind these foremen that their work has real meaning. It is important!

Control of the program must come from overall performance figures compared to similar figures from past years and by comparing one crew's performance to another's, by comparing the performance of a bucket truck crew with that of a manual crew, and comparing various size crews. We used five-man crews years ago. We now use three-man crews. We *know* we're right.

The measure of the effectiveness of a tree control program is in how well your facilities stand up under the onslaughts of nature. Recently, we suffered a major storm including several tornadoes and the most spectacular lightning any of us has ever seen. It was the most wide-spread storm we have ever experienced, causing damage over the entire 8000 square miles of our division. It continued without abatement for ten hours and kicked up sporadically for another twelve hours. It was a perfect test of our trimming program.

One hundred forty-seven trees fell on our primaries, all twisted by tornado winds. Fifty-seven large limbs fell on our primaries, all twisted by tornado winds and many carried 100 feet or more before contacting our primaries. Sixty-four limbs fell on secondaries and services. Not one single outage resulted from lack of trimming or insufficient trimming.

About every third year we are hit with a catastrophic weather phenomenon such as the storm just described or a major sleet storm. Aside from such infrequent spasms from which there is no hiding place down here, we spent about 500 company man-hours per year as a result of tree-caused, primary-line outages.

We spent $5.00 per customer per year for tree trimming.

I remind you of the instruction I gave my contractor at the outset, "Don't let the trees burn or knock down my lines." "And do it cheap."

He hasn't. He has.

So much for trees, one of the two major villains. Now for the other baddy, Lightning.

LIGHTNING

Only God can make a tree. Or lightning. Prayer is the suggested remedy. We've all done our share of praying during vicious wind and lightning storms. Nature is a formidable opponent, not to be challenged. We can only get out of its way. Remove tree growth from its path. Provide lightning with an easy access to ground. Don't resist it.

Take an honest-to-goodness sponge. Not a poly-something-or-other phoney sponge. Your sponge is bone dry. You let a drop of water fall on it. The drop of water disappears without a trace. It is soaked up completely in the dry sponge.

Take a sopping-wet sponge. Let a drop of water fall on it. The drop of water splashes all around. None of the drop of water is absorbed into the sponge.

Take a sopping-wet sponge. Wring most of the water out. Let a drop of water fall on it. Some of the drop will be absorbed instantly and some of it will splash all around.

Zero ohm ground. 500 ohm ground. 20 ohm ground.

By the time lightning gets within 30 or 40 feet of the ground it is in no mood to make alternative decisions. It wants to go to ground *right there.* Not a span away. It's up to us to provide a dry sponge—right there.

Wherever we have had repeated lightning damage we have found high resistance measurements to ground. Wherever we have improved those grounds to near zero, the lightning damage has ceased.

Lightning strikes a phase conductor at, or near, a transformer pole. The tip of the stroke is at a potential thousands of
volts above (or below) earth potential. One end of the lightning
arrester is connected to that phase. The other end is connected
to a ground wire leading to earth. The full difference of potential
is impressed across the lightning arrester and it ceases to insulate,
properly allowing the high amperage lightning current to pass
through it to ground. The high amperage lightning current sees
the arrester as a very low impedance. The arrester then, has
effectively short-circuited the phase wire to ground, so 60-cycle
current follows this path blazed by the lightning current. The
lightning arrester does its thing and snuffs the arc, but not before the lightning current has passed and dissipated itself into
the ground.

Let's say the lightning current is 4000 amps and the earth
ground measures 100 ohms. 4000 amps X 100 ohms=400,000
volts. The grounded end of the lightning arrester is now at a
potential 400,000 volts from ground potential. The voltage
drop across the arrester is about 8 KV, so the primary conductor in the vicinity is at a potential 408,000 volts from ground
potential.

If all available grounding points—the ground end of the
primary coil, the center tap of the 120/240-volt secondary coil,
the system neutral, and the transformer tank—are all interconnected, as they should be, this point of grounds interconnec-
tion is raised to a potential away from ground 400,000 volts,
leaving only the arrester drop of 8 KV impressed across the
primary coil of the transformer. This should protect the transformer from insulation spillover, coil-to-coil, coil-to-core, coil-
to-case or primary coil to secondary coil.

But the interconnected ground connection has been raised
to a potential 400,000 volts from earth potential. So what? So
has everything else nearby. The pole, the ground wires, the transformer tank, the secondary mid-point tap and the system neutral
are all at 400,000 volts. There is no *difference* of potential to
produce spillover. The primary busing, the cut-out, the primary

end of the lightning arrester are at 408,000 volts, only 8 KV away and insulated to withstand that voltage pressure. Normally the "hot" secondaries are in a range from −150 to +150 volts from earth potential. But not now. Now they're in a range from 400,150 to 399,850 volts from earth potential and insulated to withstand the differential to which they are being subjected. If there is a guy wire grounded and located too close to the secondary or neutral, we could get an arc. But we don't allow grounded guy wires that close, do we?

Other than the long-shot guy wire possibility there is no reason to expect any flashover or damage. Just like a surfboard rides a wave, these facilities will ride the voltage crest up and back down to normal.

So why fuss about high ohmic grounds? I'm coming to that. But first, let's separate that arrester ground from the other grounds. Bring a separate ground lead down the pole from the arrester and attach it to a rod located some distance from the rod connected to the system neutral, transformer ground. The two rods are not physically connected except through admittedly poorly-conducting earth. (Remember, we measured the ground at 100 ohms.)

Lightning strikes the primary. Same action. The arrester passes the 4000-amp lightning current, blocks the 60-cycle current. We have 400,000 volts at the bottom of the arrester, 8 KV across the arrester, and 408,000 volts on the primary. The transformer primary coil in series with its earth ground is subjected to 408,000 volts. If its earth ground were zero ohms the coil would get it all. We suspect that its ground will be near 100 ohms, like its buddy, the arrester ground, around on the other side of the pole. Current will flow through the coil and ground resistance and the voltage will be divided between the coil and the ground proportional to their resistances to the lightning current. The primary coil resistance, even on a small transformer, will not likely exceed about 10 ohms, so the voltage across such a coil would be only about 40,000 volts. Less on larger transformers.

So why has this separate ground connection historically produced more transformer damage than the interconnected ground connection?

First, our theory is borne out by the fact that small transformers so connected suffered many more fatalities than did large transformers. More resistance—more voltage.

Second, the transformer ground can be much better than the arrester ground for reasons having to do with the soil condition and the way the rods were driven.

Third, our 408,000 volts is a moderate example. One million volts is not uncommon.

Fourth, and most likely, the basic impulse insulation level (BIL) of the transformer may not be anywhere near 95 KV, the standard. Overloads may have damaged the coil insulation. Moisture may have deteriorated it. Previous lightning surges. Old age. Low oil level. Acidic oil.

A transformer subject to any, some, or all of these ailments is much more likely to withstand the 8 KV shock where grounds are interconnected than it is to withstand the voltage stress several or many times 8 KV where the grounds are not interconnected.

So why fuss about high ohmic grounds where the grounds are interconnected? I said I would get back to that. The answer is that whereas a good lightning arrester well protects a transformer whether the ohmic ground reading is high or low from lightning strikes on the primary, it does nothing to protect that transformer from lightning strikes on the secondary side of the transformer.

Transmission conductors gallop across the countryside for miles and miles at a height varying from 30 to 100 feet above the ground. We clear all the trees back from these conductors; trees which, if they were close, would divert a portion of the lightning strokes. The conductors are exposed and extensive. We expect them to be hit by lightning. They are.

Primary conductors gallop across the countryside for miles and miles at a height varying from 15 to 50 feet above the ground. We do our utmost to clear the lightning-diverting trees and buildings back from our primary conductors. The conductors are exposed and extensive. We expect them to be hit by lightning. They are.

Secondaries don't gallop much. Services don't gallop at all. They don't run for miles and miles. They are located 15 to 30 feet above the ground. We don't try very hard to keep them clear of trees or buildings. We don't expect them to be hit by lightning. But they are.

Transmission lines are insulated at high BIL values and are usually protected by static wires.

Primaries are insulated at a level of 95 KV and are usually not protected by static wires.

Secondaries and services are barely insulated at all and are never protected by anything. Secondary insulators have wet flashover values as low as 8 KV. I don't know what the flashover value is of a triplex service after it has rubbed a tree limb for years. Or secondary transformer terminals with a wet twig lying across them—or a dead, dead cat.

I once knew a man who bragged that he'd had 40 years' experience in utility work until someone corrected him one day while amongst an audience of his peers. "No, Tom, you haven't had 40 years of experience. You've just had one year 40 times!"

So with secondaries and services. They don't go for miles and miles, but miles and miles of them do go. Short trips, yes. But long enough for lightning to spy them and spark them.

When lightning strikes a service entering a home, it may hit a "hot" wire or the neutral. If it hits the neutral it discharges to the service ground. If it hits a "hot" wire it flashes over to the neutral and discharges into the service ground. The service ground is usually pretty good. It is connected to the water system. The ground rod is driven next to a moist foundation under a dripping eave. Voltage buildup is usually minimal.

I contend that lightning strokes on secondary or services in the first half of the span leading away from the transformer pole are the major cause of lightning problems experienced on transformer installations with interconnected or noninterconnected grounds. Lightning does not seek out the *best* ground. It goes down the *nearest* ground. In the first half span of secondary or service from a transformer pole the transformer ground is the nearest ground.

Before you say it, I'll say it. When lightning strikes the secondary system it flashes over the low-voltage insulation and the lightning current passes harmlessly to ground because 120 volts can't sustain an arc over the air spacings and insulation involved.

Right, to a point.

But let's follow that lightning current down that ground and see what happens.

We can all agree that it is much easier for the available lightning current to flash over secondary insulation than over primary insulation. So there should be more flashovers, everything else being equal (which it ain't).

When lightning strikes a "hot" secondary on service conductors in my prescribed first half span, it must flash over between the "hot" conductor and the neutral through the air, or flash over the transformer secondary bushings, or through the secondary coil. It makes no difference which, the lightning current, every bit as many amperes as flows when lightning strikes a primary or a transmission line, flows through the down lead into the transformer ground connection. There is not one brand of lightning for transmission voltages, another for primary voltages, and a wee little brand for baby bear secondary voltages.

The lightning current is dissipated into earth through the transformer ground connection. That ground connection, I have found, ranges from a fraction of an ohm to 400 ohms.

Using our previous figures of 100 ohms and 4000 amperes we find 400,000 volts potential away from ground on the butt end of the arrester, on the ground end of the primary coil, on the transformer case and on the system neutral.

So what? Where's the damage-doing difference of potential?

Well, the primary is still up there swinging 10 KV above and below earth potential. The difference of potential across the arrester and across the primary coil is somewhere between 410,000 volts and 390,000 volts. And now the arrester can't help. It is not designed to operate in reverse. We have voltage overstress coil-to-coil on the primary winding. All of the secondary coil and the ground end of the primary coil are at about 400,000 volts. The primary end of the primary coil is at, or near, ground potential. Therefore, there is voltage overstress from the primary end of the primary coil to the secondary coil. The tank and core are at 400,000 volts. There is voltage overstress between them and the primary end of the primary coil.

We can expect flashover at any one of these points of voltage overstress. As soon as flashover occurs, 60-cycle fault current will flow through the ensuing arc to ground. The transformer fuse will blow, before or after the unit is damaged, depending on the speed of the fuse, the amount of fault current and the previous condition of the transformer's guts.

When lightning strikes a primary conductor near a transformer, the arrester is subjected to overvoltage to the point of designed spillover. Thereafter, it is subject to no more than its designed IR drop of about 8 KV.

When lightning strikes a secondary or service near a transformer, the arrester is subjected to the full overvoltage I've just described, determined by the resistance of the transformer ground and the value of the lightning current. I don't know exactly what we should expect its action to be and I doubt if I'm likely to find out from the manufacturers. But I suspect that it can't be expected to work as smoothly in reverse as it does when lightning strikes the primary. It is designed for that action. Spark over and dump the lightning current into the ground connection. Can we expect it now to spark over and dump lightning current onto the primary? Why would lightning current choose to travel the primary, presumably looking for a path to ground when it has a ground connection at the foot of the pole? I doubt that it would. It seems more likely to me that this high-voltage stress would result in flashover of the porcelain and failure of the arrester. This would explain the many instances we have had of arrester failure concurrent with fuse blowing and/or transformer failure.

The only viable solution, it seems to me, is to get that ground connection resistance down to a livable level. We try to get 10 ohms or less, but we find that if we can get it as low as 20, we stop almost all of our problems.

About 8 years ago, I started a program of testing and improving all grounds on equipment poles in rural areas, by contractor. About half of my territory is on limestone. The rest has sandy, clay-type soil. If the ground measures over 10 ohms, we add a rod, and re-measure. If still over 10 ohms, we add one more rod and then give up.

Measurements are made with the down lead cut open, so we do not measure the parallel resistance of the system neutral.

We would like to use sectional rods and go deeper. We know that's the proper way. But in the limestone areas, we rejoice to get an 8-foot rod all the way down. In the sandy, clay-type soil, we haven't noticed any significant improvement by going deeper. I'm sure we would if we went deep enough, but we have had semisatisfactory results by just adding more rods, and that's easier.

We record the final reading on our system maps, right along with the height, age, and class of the pole.

Our records show that in the areas we have covered, our expense due to lightning damage has been reduced to one-fifth of the cost before grounding. I set out to cover my division in ten years. We have two years to go to finish. The savings each year have more than paid for the annual program cost. When you consider that the savings will continue forever and the costs will cease two years from now, you get some feel for the enormous benefit we will reap on out into the future.

Besides the contractor blitz, we put a meggar on every

truck involved in equipment installation. That's transformers, OCRs, primary metering, gang switches, etc. Every new ground is measured and improved, if necessary, to the maximum of three rods located 6 feet apart.

The more we work with this grounding business, the dumber we get. We are ever grateful to the really bright guys who have written all the erudite papers on the subject, but they have to base their studies on controlled conditions. The conditions we encounter are never controlled, so our findings often differ from the way it is "supposed" to be. For instance, I argued that we should concentrate on getting our ground rods down deeper instead of adding more rods. But too often we found that adding another section didn't improve our reading at all. Then adding another improved it only one or two ohms on a 60-ohm ground. Adding extra rods 6 feet away did improve the readings considerably. That's not the way it's supposed to work, but it's the way it did work, so we went with what worked and ignored theory. Not because the theory is wrong. It isn't wrong. Our conditions are wrong. Our perverse soil conditions are just not in accord with "standard" soil conditions.

We have found that our readings usually follow a pattern. They follow the automobile depreciation pattern. A car depreciates one-third each year of its last year's value. We start with one butt or driven ground and if it reads 90 ohms we find that adding a rod 6 feet away reduces that value by a third, to 60 ohms. Add another, we get 40 ohms. It isn't worth the expense to add another to get 27 ohms.

Even this pattern is broken in many cases. Our readings produce no discernible pattern. We must suspect that not all of our thousands of readings are taken properly. That may explain some weird results. But most of our weird results are produced by weird conditions.

We have also found that besides reducing our equipment damage to one-fifth, we have reduced our transformer fuse blowing to about one-fifth. And those fuses that do blow are at locations where we had to settle for less than desired ground readings. But not all.

We have installed arresters on long distribution lines where lightning damage was a regular visitation. We installed them every two miles and made sure the grounds were good. The trouble just stopped. No more trouble at all.

After properly grounding an area we always experience a rash of arrester failures. Arresters that have sat out on the line for years and years above a high resistance ground. We reduce the ground resistance to near zero and the arrester blows up during the very next lightning storm. Apparently, it had not been sparking over at all with the high ground resistance, it was old and weak, unable to snuff 60-cycle power follow current, so with the low resistance ground it did spark and the follow current finished it off. It's not in accord with any theory I've read, but we do clean all the old, bad arresters off the line immediately after grounding an area.

By now you will have decided that I am a fanatic on the subject of good grounding. I plead guilty. The evidence I have gathered over many years convinces me that I cannot afford poor grounding. If you agree with me, I have some suggestions for your consideration.

Pick out a troublesome line. Get some good ground resistance meters and measure the grounds on all equipment poles. If they all measure less than 10 ohms send me a bomb in the mail. (They won't.)

Add grounds as I've described. Maybe you'll have better luck with sectional, deep rods than I have. After you've finished, observe the results. I'm sure it will bear out my contentions.

Next, sell management on a large scale one-year test program attacking known areas of poor performance during lightning storms.

When the results of this test prove the economic imperative of a system-wide grounding program, sell that to management.

The job is too big to do overnight. Set it up over a period

of years, 10 years in my case. Approach it from two directions: a contractor to do the testing and grounding of all existing equipment poles; and company forces to test and ground all new installations they build, and your contractor forces to test and ground all new installations they build.

We use a two-man contract crew in a pick-up truck to ground existing poles. We start after the spring rains and continue until the fall rains start in order to catch the ground resistance at its worst. The number of crews depends on the funds available, of course.

The program is directed at rural areas only, because it is assumed that close spacing of equipment poles in urban areas, each protected by arresters, multiple grounding, all connected to the city water system, and the shielding provided by buildings and trees will produce less lightning problems in urban areas. That assumption is borne out by the record. However, the dramatic results we have gotten in rural areas and the number of cases of equipment failure and fuse blowing in towns (although a smaller percentage of the total) is causing us to consider a grounding program directed at urban areas if we ever finish the rural areas. It will probably be aimed at "repeater" installations rather than trying to hit every installation, many of which don't really need it.

Now we have no more problems with lightning, trees, or property owners. What else can go wrong?

FOOTNOTE

Wouldn't you know it? I have just dogmatically asserted that it is highly unlikely that any manufacturer is going to tell me if his lightning arrester runs in reverse gear as well as in forward gear. I just received a letter from McGraw-Edison confirming that arresters do not work in reverse and adding that there is no answer to lightning strikes on secondaries and services. Bless their hearts! That bears out my experience and the theory I've just outlined to you. But there *is* an answer! Low resistance grounds on your equipment poles.

POLE MAINTENANCE

Our poles keep falling over. This is not only dangerous, it is embarrassing.

In 1969 I got management's timid approval to test and treat 1000 distribution poles, all in one small area. We tested and treated at groundline with a small nonunion crew. We tested and treated above groundline with a large union crew. We "capped" the tops of the poles.

We found lots of dangerous deterioration at groundline, none above groundline. In our area of the country, our poles just don't deteriorate above ground. They dry out and weathercrack somewhat. Lightning occasionally splinters a top requiring replacement or just a haircut. But we found out immediately that the relative high cost of above-ground treatment was unjustified and that the relative low cost of groundline testing and treatment was. Since then our judgement has been borne out by the fact that we have never had to replace a groundline-treated pole which had not been treated above ground.

The method of testing and treating is well known, but I will summarize it. We test and treat 2 feet of the pole. Six inches above ground and 1½ feet below ground. We bore the pole and "read" the wood extracted. We scrape off the bad wood from the surface. The pole is then labeled treatable or reject. The rejects are further divided as "danger," requiring replacement as soon as possible, or just plain "reject" which we try to get to within a year.

We find that the poles labeled "danger" either break in two when we pull them or fall apart when dropped to the ground after being pulled. Our linemen are acutely aware of this and support the program wholeheartedly. "Reject" poles have never brought forth any complaints that our contractor is causing us to waste money replacing "good"poles.

Years ago we were all experts to one degree or another in "sounding" poles to determine whether they were bad or not. We took our hammers and banged and listened, knowingly. We know now that this was a farce and that all of the other gimmick methods are equally ineffective. Sonar, X-ray, what-have-you.

We test and treat every pole over 15 years of age. We re-treat after 10 years. We visually inspect all poles above ground for obvious defects, even those 15 years old and younger, if interspersed with older poles. Only poles in lines entirely built within the last 15 years are bypassed altogether.

Some poles are located so that it is impractical to dig all around them—poles located in a concrete sidewalk, for example. These poles are bored at a downward angle at groundline.

All poles are sounded with the trusty old hammer, from groundline to a point as high as a man can reach.

All bore holes are plugged with treated dowels.

If the pole is treatable, it is smeared from 18 inches below groundline to 6 inches above groundline with chemical and wrapped with black paper which is stapled onto the pole. The paper is just to hold the treatment in place until the hole can be refilled. It also tends to direct penetration of the chemical into the wood in preference to the surrounding soil.

The worst problem I've had with this program is getting the local company supervisors to keep proper records. They tend to let the system maps we use pile up on a file cabinet and fall off behind it. After you've mislaid one of these maps, to find which poles are danger or reject is like looking for a dime in the chicken yard. Accurate records are essential, of course.

The poles themselves are tagged to indicate the date treated and whether reject or danger.

We also ask our contractor to report any poles which are

rejected but can be continued in service by driving steel splints. This can be done for about 10 percent of replacement cost. So why don't we splint all rejects?

First, a splint looks like hell warmed over, so it behooves us to limit its use to locations not likely to produce criticism from the Little Garden Club.

Second, although the splint appears to be adequate as a pole support, we don't want to dive in head first until some of our splints have been in service several years to give unforeseen problems a chance to develop.

Third, and most determinant, is the difficulty in convincing management. No one ever questions replacement of a bad pole at a $1500 cost. You see, that's the way we've always done things. But saving a pole at a cost of $150 to splint it, strikes terror.

The cost of testing and treating is about $12 per pole at the present time. There is no way one can *not* justify it.

The acid test of the program comes when you start re-treating poles previously treated 10 years prior. On our initial round we found about 7 percent rejects. Of the poles we re-treat, we find only about 2 percent rejects. We think we have discovered the everlasting pole.

The pattern formed by the age of the rejects of poles being initially tested is interesting. 1926 poles never die. Poles treated with substitute chemicals during and shortly after World War II have a very high casualty rate. I expect they will give trouble from the ground up, even though we saved many of them at the groundline.

We include transmission poles in this program following the same schedule of treating all poles over 15 years of age. We also treat cedar transmission poles above ground and re-treat only when boring samples indicate the need. So far, this treatment is lasting at least 20 years.

Since starting this program we have carefully checked every failure of a treated pole. We have never found a case where the pole failed in the area of treatment.

Of course, we have poles broken off by errant vehicles and falling trees, but I don't recall a single pole failure due to interior decay since we completed the first round of testing and treatment about 10 years ago.

A few years prior to the start of this program we had a lineman fall with a pole. He was killed. A few years before that a pole fell with two linemen on it. They escaped serious injury. In both cases the linemen had tested the poles before climbing. They used methods then in vogue and accepted by all of us, beating the pole with a hammer and poking around the groundline of the pole with a penknife or a screwdriver.

Also prior to the start of our pole treatment program we had a pole break at the groundline and fall over in such a way that the 7200-volt conductor came to rest on the top of an automobile without touching the ground anywhere. The car had stopped at a farm gate. The driver was going to get out and open the gate. His wife was in the front seat with him and two or three kids in the back seat.

When the driver saw the pole fall and the conductor come to rest on top of his car, he had the presence of mind to stay in the car and he shouted to his wife and kids to avoid the doors, windows and door handles. He knew enough about electricity to stay put until the recloser went through its cycles to lockout. Every time the recloser reclosed, fire would fly from the door and window handles. All four tires were burned off. After the recloser quit operating, he opened the door and jumped out. He found a long, dry stick and removed the conductor from the top of his car. Then he got his family out.

Pole testing and treating is an economic imperative. It is a smashing money maker. Even if it weren't it is a moral imperative. We owe it to the men we require to climb poles and to the public who are constantly subjected to the possibility of having a pole fall on them, or an energized conductor fall on them as a result of a pole failure. If you don't buy the "moral" argument, then just imagine how much it would have cost my company if that farmer hadn't known how to protect his family from that fallen, energized conductor on his car. Easily enough to pay for our whole testing and treating program.

So much for pole treatment.

Let's talk now about preventive maintenance.

PREVENTIVE MAINTENANCE

Preventive maintenance on a utility system involves a periodic patrol of every inch of circuit. The patroler himself fixes everything that he finds wrong, if he can, and reports what he can't fix for someone else (usually a crew) to fix at a later date. He also fixes anything that, by experience, we know is likely to cause trouble, if he can, and reports it if he can't. He can and does fix almost everything he finds wrong.

This is a program which should be carried out by the serviceman. Each serviceman should spend 10 percent of his time on preventive maintenance. He should be assigned to do preventive maintenance one-half day a week, every week he is on duty. Why?

1. To insure the safety of the public.

2. To insure the safety of the employees.

3. To insure uninterrupted service.

4. To find and fix things that are wrong, or could go wrong, before they become urgent, or fail. This is much less expensive than fixing things after they become urgent or fail.

5. To keep operating personnel familiar with all the system through periodic patrol.

6. To promote good public relations. The public is favorably impressed by the sight of company trucks and

employees taking an active, detailed interest in the facilities that they depend upon for electric service. They don't need to understand what these men are doing, they just need to see them in the process of doing it.

7. To meet the requirements of periodic inspections imposed by most utility regulatory agencies. Their main aim is to insure compliance with applicable codes insofar as clearances are specified, conductor-to-ground or conductor-to-conductor.

We don't just turn the serviceman loose to wander over the system like a blind dog in a meathouse. The program must be planned.

First, the entire system should be covered once a year. Since the first round will produce a prodigious quantity of things that need to be fixed—much more than subsequent rounds—it is obvious that the extent of the first inspection must be limited. It is impossible to patrol an entire system and fix everything that needs fixing in one year if that system has not been getting such treatment in the past.

Let's look at the elments of such an inspection.

1. *Patrol the entire system, visually.* Inspecting the primaries, equipment installations (transformers, OCRs, capacitors, primary metering installations, switches, etc.), secondaries, services, service entrances (including the meter), and street lights and their circuits. Don't fix anything—just report any deficiencies observed.

This is the bare minimum acceptable. We would certainly hope that some fix-while-you're-there could be added to that bare minimum, even on the first round.

2. *The house end of the service.* The service support attached to the house may be in rotten wood, just ready to fall out. It may be located too close to the ground. It may be located in such a way that the service is dangerously close to a window.

Or so that the service is riding an eave. Now see that the drip loop is properly shaped. Unwrap the connectors at the masthead. Replace split-bolt connectors with compressions. Check compression connectors for corrosion, signs of heating. Pull on the connections to insure that they are properly compressed (goggles on). Check the customer's connection to earth ground. It may be cut, or loose, or eaten away. Pull the meter. Check and note whether the outer and inner seals are intact. Look for other evidence of tampering. Your serviceman will know, or can be trained to spot, all the ingenious devices used by current thieves. Check the meter terminals for burns or looseness. See that the meter base is firmly attached to the building. Visually inspect the service and note if it is too small, if it is old and ratty. On open-wire services note if the conductors are together or likely to get together. Note if trees need trimming or if tree guard needs to be installed. Note if service needs to be re-sagged. Tape bare connections.

3. *Fix everything which was just noted under item 2.* Or prepare an order for a crew to fix it if it is really more than a one-man job.

So far, we haven't done anything that requires climbing a pole. Only a ladder has been used. We have eliminated all deficiencies from where the service leaves the service pole to the load side of the meter. We have eliminated a large portion of our customer complaints, our preventable outages, and our unsafe conditions. At the same time we have caught a goodly number of dirty little thieves.

4. *Transformer installations.* The KVA rating should be indicated on the transformer tank in letters large enough to be read from the ground. Modern transformers are clearly marked at the factory, but sometimes the numerals get scratched or obliterated by oil and dust. Older units may never have been marked, or the markings may be obscured. A decal should be installed on such a unit or it should be stenciled with the proper KVA rating.

The cutout should be opened and the fuse link checked to make sure it is the proper ampere rating and that it is properly installed. The cutout contacts should be checked for burning, and the door or fuse barrel should make good contact when closed. The cutout should be of the proper rating. Enclosed cutouts should be checked for cracks and other damage. The stinger wires should be checked to insure that the connections to the cutout are tight. The stingers should be reshaped, if necessary, to insure proper bird clearance.

The lightning arrester should be checked visually for damage and to see that the ground expulsion device has not operated. Connections should be tightened and lead wires reshaped to insure bird-proofing. See that it is the proper voltage rating. Make sure it is tapped ahead of the tap to the cutout.

The transformer bushings should be checked for damage. Look for oil leaks around the lid and bushings. The transformer should be securely attached to the pole. Does the tank need painting? Are all primary and secondary connections tight? Is the primary tap made with a saddle? Is the hot-line clamp seated properly and is it tight? Are all stingers properly shaped?

All ground leads and connections should be checked to make sure they conform to standards. The down wire to ground along the pole should be checked to insure continuity and the connection to ground rod or rods should be inspected.

Are secondary leads sized properly?

If the transformer is obviously servicing more load than it is rated for or if it supplies secondaries that obviously feed too far, an order should be drawn to make corrections. Perhaps a recording voltmeter should be installed at the farthest customer.

A clip-on ammeter check should be made to determine transformer load at the time the installation is inspected and this reading should be interpreted in the light of known facts about when peak load can be expected.

If the installation involves crossarms, they should be inspected and an order drawn for replacement, if they are bad. The same goes for all insulators on the installation.

You can see that this is a very thorough inspection, one that consumes considerable time. Few installations require no correction. Most of the inadequacies found are of such a nature as to jeopardize service sooner or later.

We do not attempt to bring all installations into conformity with present construction standards, but we do correct any unsafe conditions.

5. *Similar climbing inspections are made at all equipment installations, capacitors, oil circuit reclosers, etc.* We check to be sure that arresters are in place. We look for "pregnant" capacitor cases. We look for oil leaks on reclosers.

6. *Tree clearance.* Near the end of the trimming cycle we will find individual cases of limbs burning in the primaries, limbs grown around equipment installations, and limbs grown into secondaries, pushing them together. These are removed by the serviceman during his P.M. rounds. Tree guard is sometimes an adequate remedy.

7. *Primary conductors.* The serviceman can make some repairs to primary conductors, but most repairs are too much for one man so he makes out an order for a crew to follow up. What he finds are bad splices (hand-twists, patent splices, poorly made splices, too many splices in one area indicating that the conductor may have been damaged during storms, repaired temporarily and forgotten), splitbolt connectors, poorly sagged conductors, obviously inadequately sized conductors and old nonstandard conductors, such as #4 ACSR which has a history of giving trouble. He will also find places where more than one size conductor has been spliced together.

8. *Secondary and street light conductors.* Here the serviceman will find all of the same problems he finds on primary conductors, plus obviously over-extended secondaries. Over-extended secondaries call for installation of recording voltmeters and preparation of an order to add a transformer, or enlarge the secondaries, or both.

9. *General repairs.* The serviceman may recommend that a pole be replaced due to above-ground deterioration. He may

recommend that a pole be added in a span that is too long. He may recommend that a pole be replaced to provide adequate ground clearance or clearance to telephone or CATV conductors. If he finds a broken insulator in a secondary rack he may replace it himself, or prepare an order for its replacement. He will do likewise for pin-type primary insulators. For broken primary dead-end insulators he will usually have to prepare an order for a crew to replace. He can tighten down guys and install guy guards, but will probably have to prepare an order to get a span guy repaired. He can usually tighten loose hardware himself. Bad crossarms must be replaced by a crew. He can install tags on tap line poles to indicate the proper fuse link size to be used in the tap line cutout. He can replace broken street-light glassware.

Now the serviceman has done just about all that we can expect from a single lineman working without any help. There are some aspects of preventive maintenance which are best undertaken by a small crew equipped with a bucket truck.

10. On a hot day in summer, after a series of hot days, it is a good idea to send out several small crew units to make *clip-on ammeter checks of suspect transformers* and to "clip" the primary circuit current at strategic locations. Switch maintenance and recloser maintenance is a job for a crew. Of course a crew is used to follow up on all the orders generated by the serviceman on his P.M. rounds.

After covering the entire system each year for three or four years, expanding the coverage a little each year, we will find that we have reached the point where we are finally repeating ourselves. You would think that at this point everything would be in good shape and that we could no longer find anything to be fixed. Well, of course, it doesn't work out that way at all. We have had this program going for over 20 years and we are still finding split bolt connectors. Also, the physical system continues to age and deteriorate. Telephone and CATV companies continue to install new cables creating new clearance problems. Trees keep growing. Anchors give, allowing lines to sag. On and on. We will never run out of things that need fixing.

I said at the outset that this program cannot be turned loose to run itself. It must be planned, supervised, and proper records are essential. Now that the essentials of the program have been described, we are in a better position to discuss how it should be planned.

As I pointed out, it is impossible to completely cover a system the first time in one year, and do everything in the way of repairs and improvements that should be done. I also specified that each serviceman should spend 10 percent of his time on preventive maintenance. Now that sets the limit of what it is possible to do. Some offices have only one serviceman, some two, some several. How many servicemen (linemen) are available in a particular office determines the number of man-hours available to perform preventive maintenance in the area served by that office.

We now have a known area, we know its physical condition, and we know the manpower available to go over that entire area in one year's time. The areas may vary considerably in size and the condition of the facilities may be much worse, or much better in that area than the average. All of these factors must be taken into consideration when deciding how deep to dig in the first round. After the first month it can be decided whether to go deeper or lighten up in the remaining 11 months. But the entire area *must* be covered in the course of 12 months.

Item #1 (visual inspection, only) on my list of elements of inspection is the bare minimum, as I said. Most situations will allow for item #2 to be included, also, in the first round. (That was the service entrance.) You may even get in a little of the "fixing" called for under item #3. You're not likely to get into items #4-9. You can start on the crew item #10.

The second year you will start with a lot more confidence in what should be accomplished by one serviceman in one half-day run on preventive maintenance. On this second round we repeat item #1, but we can bypass item #2, except for a cursory visual inspection to catch obvious deficiencies like house knobs pulled out, or services together in trees or rubbing against tree limbs. Now we concentrate on item #3 and move into item #4, transformer installations. Yes, we're now involving the serviceman in time-consuming detail work. But on the second round, item #1 does not require nearly as long because it was "cleaned" on the first round. Also, the serviceman is now becoming very familiar with the system and can inspect it much faster. He has also developed inspection techniques which allow him to move rapidly.

On the third round we repeat item #1, skip items #2, 3, and 4, if they are complete, and move on to items #5-8. Item # 6, tree clearance, gets particular emphasis on this round, whereas up to now we have just attended to those tree conditions that qualified as emergencies. This time we begin to get our services and secondaries cleaned up. After this round we can expect to have no outages from this cause unless we have a powerful wind storm.

On the fourth round, we repeat item #1 and concentrate on item #9, general repairs. We also drop back through items #2-8 on a selective basis to get skips from previous years and to inspect new facilities installed since inspection began.

By the time the fifth inspection rolls around, we should be ready to start all over again. This time it won't take four years to work our way through the items. By now, we should be able to set up a schedule in which we cover all 10 items in two years.

But you have to be able to *prove* that you went through all 10 items in two years. So you need to keep records.

Each serviceman should be given a map of the area for which he will be assigned responsiblity for the year. He should outline the area he has "worked" each day, write his initials in the center of that area and the date. His supervisor must check this map once a month to insure satisfactory progress.

Each serviceman should be given a form on which are listed down the left-hand side of the page all of the preventive maintenance units of work. Units of work include such items as "Primary spans, trimmed"; "Ground wire repaired"; "Service resagged"; "Meter base tightened"; etc. His units are totaled at the end of each month and submitted on a composite report to the division office, showing the results of each serviceman's efforts for the month.

At least once a quarter the superintendent of construction and maintenance should make a spot check on each serviceman's work. He should pick out a recent day's work to inspect. He should take the serviceman over the area worked and check the items reported. Now it is important that this not be done in a spirit of trying to catch the serviceman in a lie. Rather, it should be done in the spirit of evidencing a lively interest in the good work being done by the serviceman. Now, the serviceman is not stupid. He knows the superintendent is checking up on him. And that's not bad. His work should be checked, and he knows it. But it has been my experience that the serviceman, rather than resenting this inspection of his work and fearing that he is going to get caught short in the pants, welcomes the superintendent's visit and is eager to show him all the horrors he has found and corrected. He takes a proprietary interest in "his" area and is vitally interested in not having anything go wrong in "his" area as a result of his missing something on his P.M. inspection.

To promote this feeling on the part of the servicemen, a district-wide meeting is held once a year near the end of the P.M. cycle which is attended by the district manager, the district superintendent, the local managers and the servicemen. At this meeting plans are made for the next P.M. cycle. The servicemen are in the best position to speak from experience and their suggestions usually predominate in deciding what tack to take for the coming year. We never let them forget that it is "their" program.

Each month the serviceman's P.M. hours are credited to his efficiency rating at 100 percent. The percent of the district's serviceman hours is calculated at the division office and reported back to the district along with the serviceman group E/R rating.

I doubt if I need to belabor the value of this program in terms of improved service quality and continuity. If you have any experience at all in electric utility work, you know that this type of work is bound to produce bounteous dividends.

It is difficult to quantify the value of preventive maintenance, because we have involved ourselves in so many system improvement programs in the last 25 years that it is impossible to measure precisely the exact effect of each one.

We did have a measure of the effectiveness of preventive maintenance when we first started because we started only in one district. It took years to get all of our districts to join in and some still lag behind. The improvement in service in that first district to start P.M. was dramatic. Outages were cut to a fraction of the previous experience. Quality of service in that district was far better than in the other districts which did not have the program underway. Then as other districts took up the program their service improved markedly. The more P.M. performed in a district, the less overtime we experienced.

Comparison of my division with other divisions shows that our overtime per customer is less than half of the next best division and less than one-fourth of the worst division. Much of this record must be attributed to our P.M. program. P.M. is a money-making son-of-a-gun.

INSPECTIONS

Periodic Inspection of substations, reclosers, and capacitors is an essential part of preventive maintenance. But we do not include these inspections in our formal P.M. program for several reasons.

P.M. is accomplished by a lineman–serviceman who devotes a minimum of one-half day, uninterrupted, to system inspection and repair, using his own judgment in routing himself and in determining how far he will go in making needed repairs. The periodic inspection of substations, reclosers, and capacitors does not fit into this type of activity. It allows for no such independent action and judgment. The actions involved are clearly defined and there is a report form for each of the three types of inspection. Each such inspection is a work unit under the efficiency rating system, so the serviceman is credited on a standard unit time basis for inspections while being credited actual hours under the formal P.M. program.

When I first suggested that local servicemen be required to go into the area substations periodically to inspect them, I thought my substation superintendent was going to suffer heart failure. He was very proud of his stations and just could not visualize allowing the local boobs to enter his golden sanctuaries. Most of them had never been inside a substation fence for any purpose other than switching. They were not allowed to touch a regulator.

Reluctantly, the substation superintendent agreed to arrange training for all of our servicemen. As soon as he got into the training he began to realize that these boobs could be a lot of help to him, that they were very concerned about the proper operation of their stations and were not only eager to learn, but were very capable of learning.

He took these men through the station and pointed out how the transmission voltage enters the station, is transformed to distribution voltage, how the distribution is divided into circuits, each usually protected by an automatic breaker or recloser. He explained the function of the transfer bus and showed them how to bypass a breaker and test it. He explained all the gauges and dials on the various pieces of equipment and familarized them with station metering, explaining the C.T.s and P.T.s. He pointed out the importance of keeping the structure free of birds' nests which invite invasion by snakes. He emphasized the importance of reporting any rust on equipment, structure, or fence and the importance of keeping lonesome weeds pulled. They learned how important it is to report any gaps that appear between the rocked floor of the station and the chain-link fence. He urged them to keep trash picked up inside the fence and around the station entrance. He showed them how to record breaker/recloser operations since the last inspection, explaining the importance of recording phase and ground targets accurately. He explained the operation of the regulator and how to reset the position indicator and how to record that information. He used an ammertong to show them how to take current readings on each phase of each circuit.

A simple form was devised on which to report the pertinent information gleaned from each monthly inspection.

A serviceman visits each line recloser installation once a month and records the counter readings on a report form, a copy of which is filed in the district office and a copy in the division office. If a recloser shows too many operations, a line patrol is called for to determine the cause. This situation often turns up a bad lightning arrester or a tree burning on the line, or phases slapping together in a long span while a cow rubs her haunch on a down guy. If there are no recorded operations, even after a bad storm, we suspect that the unit is malfunctioning and inspect it and test it.

Each capacitor installation is visited by a serviceman once a month. He checks for failed units or blown fuses. He records counter reading on switched banks. If there are too few or too many operations, the controls are suspect and must be tested. If there have been no operations, on current or KVAR controls the bank may be located at a point where there is insufficient power flow to actuate the controls. It may have to be moved. And the engineer who specified that location will have some explaining to do.

SYSTEM COORDINATION

It would be presumptuous to go into a detailed explanation of system coordination because there is an abundance of literature on the subject written by folks who know lots more about it than I do. But its importance to the efficient operation of an electric utility system demands some mention. A discussion of some of the problems which arise *after* your engineers are fully trained in the scientific aspects of this program is in order.

First of all, I have found this program produces more foot-dragging at all levels than any other I've attempted. I held classes and fully trained my graduate engineers in the technical details of "doing" a system coordination. I naively expected them to charge forth and spend perhaps too much time on this purely engineering activity. After all, hadn't they always complained that we didn't give them enough opportunity to utilize their engineering education?

After repeated urgings, with no effect, I finally instructed them to choose one day a week on which they would isolate themselves and spend the entire time, uninterrupted, on system coordination. I have never heard so many ingenious excuses. I concluded that an engineer just can't stand his own company for a whole day. At any rate, I got precious little coordination done.

I always resisted making specialists out of my engineers. I wanted them all to get the full dose so they would be better prepared for any opportunities which might come their way, or any service I might see fit to put them to. But I had to quit riding that dead mule after years of disappointment. Most engineers just don't like to perform purely engineering basics. They thought they put that behind them when they graduated from engineering school and are genuinely hurt when you press them to immerse themselves once again in prolonged scientific thought.

So I specialized. I found one engineer who likes to spend his time in scientific calculation and turned the whole division's system coordination over to him. And he does it with a vengeance.

But after he completes a system and makes his recommendations we run into some more foot-dragging. Here we have a superintendent who has complained bitterly that one of his substations and its lines should be properly coordinated. We finally get to it and report the results to him along with recommendations for relay settings, added reclosers, relocated reclosers, changed fuse link sizes and added cutout locations.

Is this longed-for information greeted with hallelujahs? It is not. Progress toward implementing these solutions is analagous to wading through a hogwallow—barefoot.

You may be wondering why I don't just order these engineers and superintendents to do what I want them to do. Well, did you ever try to push on a rope? I have had more success pulling than pushing. For instance, when one of my daughters was 5 years old she refused, after repeated urgings and explanations of the efficacy of proper hygiene, to wash behind her ears. It finally dawned on me that this 5-year-old girl's ears were *supposed* to be dirty. Now that discovery delivered both her and me from an extended period of mutual misery. When she reached the age where a little girl's ears are *supposed* to be clean, they got clean.

When the time gets right for an electric utility system coordination recommendation to be implemented, it is. You just have to grit your teeth and be patient. It always gets done. And faster if you don't nag.

Now, you won't read any of the foregoing in a G.E. system coordination handbook. But it is just as vital to getting from thinking about coordination to having coordination as any of the good stuff you'll read in a handbook.

Two things stand in the way of proper coordination which are particularly aggravating. One is old-type relays with their flat curves. It would be nice to have all new-type relays with their curves standing up parallel to those modern fuse curves, but it's too expensive to throw away all those old dogs and substitute new ones. We try to move these units to small stations where they produce less conflict with high side fuse curves. We also try to replace a few every year so as to eventually work most of them out of the system.

The other snag is high resistance substation ground. All the calculations in the world won't make a coordination scheme work if the station ground is so lousy it blocks the return of fault current.

This brings up some sore points. First, how do you get an accurate measurement of station ground? And second, when you find that it is lousy, what do you do about it?

We need two measurements of station ground. One is the resistance seen by lightning when it strikes the station, or very near the station. We are told that from zero up to about 10 ohms this resistance is equal to the resistance we read with a meggar device. But from 10 ohms up, it is less than that. How much less I have been unable to find. It doesn't make too much difference since our station ground must be close to zero to assure proper operation when lightning strikes. We sit on a bed of limestone, so we have stations with as much as 5 ohms resistance which operate semisatisfactorily.

The other resistance is the resistance seen by fault current trying to find its way back to the station, to the grounded common point of the Wye. Again this should be near zero, and again we have stations where 5 ohms is the best we can do.

The measurement of this resistance is the subject of much controversy and a lot of scholarly papers. The problem, it seems to me, is that the instruments and their leads are so sensitive and so subject to outside influences that the results are questionable.

To measure station resistance to lightning current we should disconnect all connections to ground and measure from a point on the ground grid only. To measure resistance to fault current we should measure from the common Wye point with all connections to ground and neutrals intact, since all these paths are available for return of fault current.

It seems to me that we should keep in mind the purposes of these measurements. First, we want to know what voltage rise we can expect at the station as a result of the flow of lightning current. And we want to know how much voltage drop we will experience from the point of a ground fault back to the station. If we could devise a zero resistance return path for the fault current, all of the phase voltage would be dissipated from the station out along the phase wire to the point of the ground fault and the voltage along the return path would be zero. If the impedance of the return path were equal to the impedance of the phase wire, the voltage dissipated along the return path would be equal to one-half phase voltage.

Second, we want to know how much fault current we can count on to flow if an energized conductor comes into contact with a grounded surface.

It seems to me that it is a better method to actually ground a conductor and measure the flow of fault current, than to pursue the tenuous and questionable methods available to measure the ground resistance by applying a tiny voltage to push a tiny current over thousands of feet of lead wire subject to innumerable stray influences of inductance and capacitance, thence back through this ground resistance. Consider this possibility with me. Insulate all neutral returns and statics, except one. Tie all of the insulated returns to this one return. Drive a ground rod at a point on one circuit 3000 feet from the station and another one 6000 feet from the station. Connect a ground lead to each rod. Mount a temporary cutout on each of these two poles and attach one terminal to the ground lead and one to the primary. Rent two ammeters which will read and record fast currents and attach one on the station ground lead just before it enters the Wye point and the other just on the station side of the confluence of all the neutrals and statics. Close the cutout at the 3000 foot point and record the two currents. Then close the cutout at the 6000 foot point and record the two currents.

You now have four current readings. For the 3000 foot length, you have the total fault current which returned to the transformer bank. You have the portion of that current which returned through the multigrounded neutral system. The differences between the two readings is the current which returned through earth only. You have similar values for the 6000 foot length.

We now divide the total current into the single phase to ground voltage, as read at the time of the test. The result is an impedance value. This impedance value is made up of three impedances in series: (1) the impedance of the transformer; (2) the impedance of the phase conductor from the station to the point of the fault; (3) the impedance of the return path. The return path is made up of two impedances in parallel: (1) the impedance of the return path from the point of the fault via the multigrounded neutral system, terminating at the station at a point where we have collected all neutral returns into one. We have applied a meter there. Past the point of the meter this path dips down to the station ground grid. (2) The impedance of the return path from the point of the fault via the earth, terminating at the station ground grid, where it is joined by the multigrounded neutral return, thence up the ground connections to the single conductor finally leading into the point of the Wye. We have installed a meter there.

We would now like to know the voltage drop from the point of the fault across the two return paths, in parallel, back to the station. This is the voltage that is left after we calculate the voltage drop from the station through the transformer, thence along the phase conductor out to the point of the fault and subtract it from the applied voltage. We know the impedance of the transformer and the phase conductor and the current through them, so we can easily arrive at that figure by multiplying it by the total fault current read. Subtract it from the applied voltage (vectorially, of course). The voltage you have left is the voltage drop across the combined return paths. This voltage, divided by the total fault current, gives us the impedance of the combined return path. Since we know the current in each of the two return paths, we can calculate the impedance of each path, one through the earth and the other through the neutral system.

The earth impedance is the one seen by lightning and the combined impedance is the one seen by ground faults.

You will recall that we made our test at a point 3000 feet from the station and another at 6000 feet from the station. Why? Because we want to assure ourselves that we are reading the ground resistance from a point referred to as "Remote Earth;" that is, a point at which the resistance back to the station grid will not be noticeably exceeded by any measurement taken farther from the station.

Let's look at that concept a minute. If we measure the ground resistance at a point zero feet from the station ground grid, it would be zero ohms. Move away 10 feet and it would increase to some tiny figure, then at 100 feet still more, and so on until at some point it would cease to increase. Now, of course, it wouldn't be all that smooth. You would get some surprising variations along the way, but it would follow that general pattern. So if our earth return path resistance at 3000 feet and 6000 feet are essentially equal, we can assume that we reached "remote earth" at 3000 feet. If 6000 feet is, say, 15 percent higher than 3000 feet then you can accept 6000 feet as "remote earth." If 6000 feet is twice as high as 3000 feet, then you'll need to make another test at about 10,000 feet, and so on until your readings cease to change appreciably.

What effect can we expect the ground resistance (measured by meggar at the point of the fault) to have on our test? Obviously, we want to select a pole where the ground resistance is as nearly zero as possible, so our results will be contaminated with as little extraneous influence as possible. If we have chosen a pole to fault and find that the ground resistance is high, we ought to lower it, or move to a more congenial location. A rod should be driven at the 3000 foot point and at the 6000 foot point finally chosen and it should be meggared before connecting it to the pole down wire. But don't be surprised if the test result is less than this meggar reading. The meggar only reads the resistance of a path through earth from a point about 60 feet away from the pole, in one direction only. When we apply 7200 volts at this point, the resulting current is going to find every available path through the earth back toward the station and there very well may be a much lower resistance seam of earth in the vicinity than any our little bitty meggar current found.

What we want to avoid is locating our fault at a point of high localized ground resistance which is not representative of the localized ground resistance to be found generally in the area. The localized ground resistance found generally in the area is certainly an effective portion of the resistance of the earth path back to the station, and of the neutral path back to the station. That's the kind of soil a phase is going to fall on if a phase wire falls. That is representative of the kind of soil into which lightning current must be absorbed if lightning strikes the station. But you wouldn't insert your ground rod into a projecting lonesome boulder and apply your fault at that point and expect to get a current which would yield calculated ground impedance indicative of values normal for the area.

Now let's go back and look at that business about insulating all neutrals but one, and tying them all together so that all of the neutral return current is forced to flow through our meter. That is a helluva lot of trouble, physically. It just isn't practical if we're going to get the most out of our meters while we have them rented. We want to get as many station impedances measured as possible, so we can't complicate those measurements any more than necessary.

Look at the point of the fault. Current is flowing through our two return paths back to the station, producing a voltage drop along its length. This drop is essentially the same through both the ground path and the neutral path throughout its length, foot by foot, so there is little inclination for the current in the earth to ride the neutral and vice versa. Point by point along the route there is no difference of potential between the two paths, for all practical purposes.

The circuit being short-circuited leaves the station and travels, let's say, in a northerly direction. The other circuits will be travelling East, West, South, etc., diverging more and more the farther you travel out from the station. There is little impetus for our fault currents to go wandering off to hop on a neutral concurrent with another circuit to find its way home to the station. The difference of potential which might cause this is greatest near the point of the fault. At that point the impedance to alternate return paths of other circuits is greatest, too. Both the difference of potential and the impedance between circuit paths diminishes as we near the station. From this we can see that there will be little inclination for fault current returning along our neutral path or along our earth path to wander off to any appreciable degree and show up at the station in neutral dead-ends associated with other circuits. And it doesn't.

Just to be sure that some peculiar local soil condition doesn't force the earth current back through an alternate circuit neutral, we move our C.T. to that alternate circuit neutral and blow another fuse and measure the resulting return current. It will usually be a very small percentage of the total earth return current.

What does this small amount of earth current returning by alternate circuit neutrals tell us? It is an indication of the degree to which the station grid may be considered to be extended by these alternate neutrals with their connections to ground. They "help" the station grid absorb lightning current to just that extent. They are included in calculating our earth resistance and reduce it by just the proper amount.

Therefore, we choose to take our neutrals as we find them. Place a split-core current on the neutral of the circuit being faulted at the point where it hits the station. Place another on the neutral wire leading to the Wye point. The currents that we read will be sufficiently accurate for our purposes without bollixing our test set-up with a lot of unnecessary and ineffective complications.

If this experiment reveals a station resistance so high as to be unbearable (I'd say anything over 2 ohms) my suggestion is to drill a well and drop a 4/0 copper wire down it. I don't think driving more 8-foot rods at ground level will do any appreciable good. You need to reach down to water table. And that means water table during a drought, so don't skimp on depth.

After doing what you can to improve the station ground, it must be remeasured in the same manner I have just described.

All of this testing costs money. We don't do it just to get our jollies. It's worth far more than the cost. How many instances I have observed of trouble reports describing phase wires on the ground burning, and nothing happening in the way of protective equipment operating to clear the fault. Think of the danger to life! Usually, we're lucky. No one touches the fallen, energized conductor. But what defense would the company put forth if someone did? Often such instances result in fires being started. One good lawsuit would pay for all of our testing many times over. But beyond that, it is unconscionable to abide such a situation. We are entrusted with the responsibility to keep these wires up in the air, if at all possible, and to see that they are de-energized when they fall.

Another aggravation is trying to coordinate high side substation fuses with transmission relays where the fuse sizes are dictated by large size substation transformers, such as 20 MVA transformers. Sometimes, it just can't be done. The transmission wire size dictates low relay settings to protect against transmission wire burndown. In this situation, you can compromise as best you can and look the other way.

I like to see tap line fuse link sizes posted on the cutout pole. This can be done by tacking aluminum numerals on the cutout–tap take–off pole, or by taking a plastic sheet with the proper numeral printed on it and tacking it on the tap pole.

Sectionalizers are a pain in the neck. We still have a few in operation and won't remove them until they cause too much trouble. It's a grand idea on paper, but it don't work in practice. The problem is that if you have enough sectionalizers and reclosers out there on the system, all backed up by a larger recloser or breaker, a general lightning storm will inflict so many strikes on the system as the storm front passes through that the back-up recloser will be "counted out" and lock out unnecessarily before any one of the sectionalizers or smaller reclosers goes to lockout. The purpose of installing the sectionalizers has been defeated. Instead of having no customers out of service, or a minimum number, you have a major segment of your system out. This is not only true of sectionalizers. You can experience this annoyance if you have no sectionalizers but a lot of reclosers located downstream from a back-up recloser and located too close to that back-up unit where they are all subjected to multiple lightning strokes from the same storm. The reason that sectionalizers are particularly worrisome is that they are cheaper than reclosers and you are inclined to use more of them (often where a fuse would be adequate), laying yourself open to this "count-out" phenomenon.

Records. Here I will repeat myself because it is so important to keep good records and to "read" them every month and take appropriate action.

Substation relay records must be recorded on each monthly inspection, showing all operations, phase, ground, and instantaneous. Recloser operations must be recorded as completely as the recording device allows. Whenever relay settings are revised, everyone involved must be duly notified in writing.

Spike Jones used to sing that there were nine buttons on her nightgown, but she could only fascinate. There are 900 parts to a recloser. When it is dismantled for maintenance and then reassembled using only 899 parts, it won't work right the next time there is a fault on the line. Maybe all the parts are put back in the thing, but one part is reversed. The point is, that the finest system coordination program goes blah without expert maintenance of reclosers and breakers. *Real* expert maintenance men go by the manufacturer's maintenance manual like it was inspired scripture.

OCR MAINTENANCE

Substation breakers and reclosers should be maintained on a schedule as specified by your standards and the work should be done by substation maintenance specialists. Line reclosers ain't so lucky. Every two years they should be removed from the pole, returned to the shop and maintained according to the manual. This is done by construction personnel. At least one in the group should have had special training so he can keep the others from botching the things up.

Often overlooked is the event of enlarging a substation transformer. This increases the available fault currents and calls for recoordination of the system. Most of the effect peters out after you get over half a mile from the station, but it can subject equipment to fault currents in excess of specified rating, and it can foul up coordination between devices operated in series. All very obvious. But how often through the years I've had a system coordination breakdown which was traced to just this cause!

SWITCH MAINTENANCE

It promised to be one of the biggest fires the little town had ever seen! Why is it that disasters bring out the worst in all of us? Don't tell me it doesn't give you a thrill to see someone else's house on fire. "On-the-scene" TV cameras are glued to the scantily clad young lady trapped on the fifth floor of a burning tenement. Don't tell me you turn the set off. No, you let supper get cold while you watch.

But it's not so much fun watching your own house burn down. That's why I had mixed emotions when the old feed mill went up and I noticed the flames moving in the direction of our transmission line which passed nearby. Like watching your mother-in-law drive over a cliff in your new Cadillac.

I went with a serviceman to the nearby switching station from which we could see the flames moving ever closer to the transmission conductors and an inevitable city-wide electrical outage—including the fire pumps! All we had to do was switch that section of the line out of service. No loads involved. No one would be out of service. Just open this switch. The switch on the other end of the line segment was already open. Just simply open the switch!

The serviceman removed the lock from the switch handle, raised the handle to horizontal, and with both hands in position braced himself for the usual rapid switching motion. He gave a mighty heave. His guts darn near dropped out. It didn't budge. Now you know how this causes one to react. It is an insult for a man to pit his mature masculine strength and operating wisdom against a 1½-inch piece of pipe and have the pipe laugh at him.

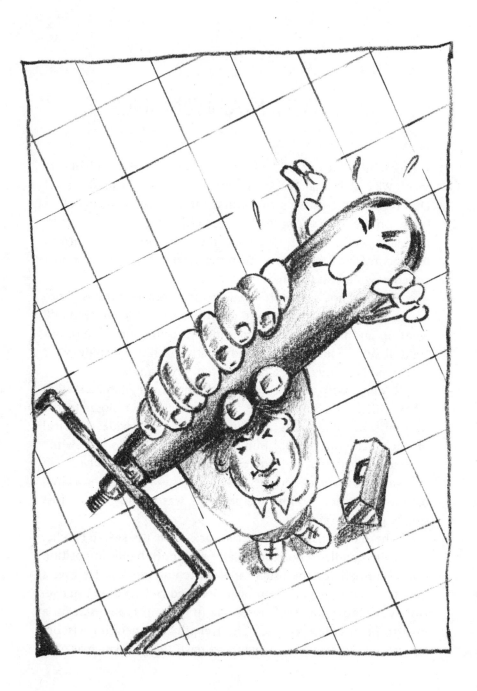

So he gave it another magnificent heave—and another. Then he shined his light on the switch. The switch just quietly chuckled at him. Once more! Heave! . . . . . nothing.

The flames were licking the aluminum transmission conductors.

The serviceman disgustedly described the switch to me in uncomplimentary terms and reported on the radio that the switch was frozen.

I should finish the story and tell you how we worked our way out of this mess, but I won't. I've already made my point. Switch maintenance. This switch hadn't been operated for three years. The next day a lineman in a bucket tapped the contact end of the switch arm and it flew open. He inspected it and found a small amount of corrosion on the contacts. But just enough to freeze it fast.

Switches are mighty handy for the efficient operation of your electric system. But if they won't work, they are a curse.

They can also be dangerous. A switch that is not maintained can fall apart on you when you operate it. With spectacular results.

What about the switch that closes—all but one phase?

Every switch should be operated at least once a year and observed for smooth operation through its cycle. The contacts should be greased and it should be adjusted for proper horizontal and vertical alignment.

A switch, like a good wife, is easy to neglect, with similar disastrous results.

CONTRACTORS

Bureaucracy is a mental illness. The disease is by no means confined to government employees. It just tends to thrive in an environment, such as governmental offices, the way Bubonic Plague tends to thrive in an environment of rats and filth. Totally free enterprise will not support the existence or growth of this deadly virus. The farther business moves away from free enterprise, the more it becomes subsidized and protected by the police powers of government, the more we see it affected by the cancer of bureaucracy.

Until one comes to understand that the aims of bureaucracy are 180° out of phase with those of free men, one is continually befuddled and confused by the apparently senseless actions of those both in government and in business who are afflicted with the bureaucratic malady.

One brilliantly clear aim of free men in business is to accomplish the most with the least manpower. A bureaucrat does his utmost to expand manpower and progress toward the conclusion of any task at the most maddeningly slow pace possible. In this way he supports expanding his manpower to maximum limit. The more manpower he can corral under his supervision, the more important his job appears and the more salary he can command. This justifies a bigger desk, a bigger office, another secretary, etc., etc. In a word, it puts more *power* in his hands. And power is the name of the bureaucratic game.

On the other hand is the picture of your local millionaire home building contractor who drives a beat-up pickup truck, wears work clothes (even on Sunday) and handles all correspondence via hand-scribbled notes on the backs of old envelopes.

He operates with a mimimum work force and his motivation is to see buildings built as efficiently as possible. His concern is to beat the competition by providing the best product at the lowest cost. He knows that unless he does that he'll be out of business.

The only competition the bureaucrat has is other bureaucrats whom he must outdo in being inefficient.

Public utilities fall somewhere in between these two extremes. We used to brag that we were privately-owned tax-paying businesses who had to provide superior service at low rates and still make a decent profit for our owners. We structured our own rates and simply submitted them to the state commision for review and approval.

Today, we no longer mention private, tax-paying, or profit. We emphasize "Public." We don't really have much to do with setting the price for our product. First, the state commission took over that function and now the federal government has just about run the state out of the field. We are in the position of the Arab caught stealing who is forced to watch as they cut off his hand. I saw a series of pictures of that operation years ago in a magazine. What impressed me was the stupid smile on the thief's face as he gazed at the stub.

Utilities today are being forced to watch as various arrogant governmental agencies whittle away at our capability to provide vital electric service to our customers. Our primary purpose is to be in "compliance." The majority of our supervisory talents are absorbed in this effort. Less and less of our attention can be directed toward our old boss, the customer, while we eagerly grovel and whine in a vain attempt to please and appease the naked power of government.

On a scale of one to ten, with the thorough-going bureaucrat at zero and the free man at ten, utilities are now at about 1.3.

When I entered the utility business over 37 years ago we were at a level of about 8. I have seen a steady deterioration which could be represented by a straight line tilted down, running left to right, on logarithmic paper.

Things are rotten, sure, but I don't recommend surrender. We can only hope that the American people will wake up to this threat and turn the scoundrels out, with our help in our roles as private citizens. In the meantime, it is our duty to stand by what we know to be right and persevere in our attempts to run our business as nearly like free men would as we can.

The most serious threat to that position is the increasing tendency toward bureaucracy at all levels in our own business that inevitably goes with constant contact with and interference from bureaucrats. This manifests itself in the consistent, growing pressure to add employees. Control of the payroll is *the* control over the operation of any company.

Whenever an additional employee is added to the payroll you take on lots more than just the expense of his paycheck. There are the fringe benefits, added vehicles, supervision, office and storage space, and increased incidental expenses. But the real threat to an efficient operation is size itself. Whenever you allow the size of your organization to pass a certain point, you lose not only the practicability of control, but also the desire for control.

If my organization were large enough to perform with our own employees all of the functions we now carry out with our own employees *and* the employees of the contractors we hire, I would not be able to exercise a fraction of the control I now maintain over both company employees *and* contractors.

Each contractor we utilize requires only one segment of my control activity. If all of those contractors' employees were *my* employees, each one of those employees would require one segment of my control activity. There just wouldn't be enough to go around. You see this sorry effect in the operation of government at all levels and you see it in the operation of all large corporations. A utility is in a unique position to operate as a lot of small units, spread out over a large area, as we usually are. Utilities serving one large cosmopolitan area—sorry—you're stuck with the problem I'm talking about.

Another reason I couldn't maintain control over all those

employees is that management wouldn't let me. That large an organization is too juicy a morsel not to bite into. It would be divided and put under several supervisors. And they would be right. One man could not stretch himself out that far.

I constantly look for ways to reduce manpower by utilizing more efficient tools, equipment, and methods.

The most readily available method is to contract as much work as possible. What kinds of work should be contracted?

Dirty work. Work that is beneath the capabilities of the well-educated employees you want on your payroll. Stupid work. Work that requires only limited intelligence. Seasonal work. Work that is required during certain seasons of the year or is best accomplished during limited periods during the year. Occasional work. Work that arises at odd times and requires significant manpower for only a limited time. Specialized work. Work that requires specialized talents or equipment which we cannot support on a full-time basis. Surplus work. Construction work in excess of that needed to occupy company forces during the slowest month of the slowest year. Major transmission line construction.

Some contracted work fits more than one of these descriptions.

We contract almost all transmission line construction, whether large bid jobs or smaller jobs handled on a unit price basis. We do not want to employ manpower and purchase equipment which would be impossible to utilize on a continuing basis.

We contract all of our distribution line work over and above that which we can count on as a continuing minimum. In this way we avoid expanding and contracting our manpower and equipment with fluctuating requirements. We are never faced with "slack" times when our people are forced to loaf, or work on marginally productive projects. Company forces should always be kept busy. "Idle hands the Devil's workshop make."

Also, no company work force can ever start a sizeable project and pursue it to completion without being called off for an emergency. This is very inefficient and the job is left for long periods of time in an unsafe condition.

Add to that the peculiarity that contract crews seem to be able to prosecute difficult, complicated projects safely which company forces approach with the utmost timidity.

The substation force must stay away from home much of the time to cover our 8000 square mile area, so we try to minimize the portion of station construction and maintenance which we do not want to trust to contract help. Therefore, we contract grading, fencing, installation of ground grid, road building, blacktopping, foundations, weed control inside the fence, mowing outside the fence, and some steel erection. When we get really pushed we let a contractor install all the steel, busses and switches, but only with a company inspector on-site at all times.

In connection with substation work, we contract all large equipment hauling and wherever a crane is needed for lifting large equipment we use a contractor.

Ditching for underground installations is contracted to local contractors.

A contractor repairs and reconditions our distribution transformers.

Our incandescent street lamps are replaced and cleaned on a group basis every nine months and our mercury lamps are replaced on a group basis every fifth round by a contractor.

Our change out meters are hauled by contractor to our meter lab for periodic tests. This used to be done by local servicemen who piled the wife into the service truck and made a shoping day of it once a month.

Of course, we use contractors to maintain tree control, whether trimming distribution lines or applying chemical to transmission or distribution lines. All initial clearing and bulldozing is done by contractor.

I've already discussed pole and anchor installation by contract. Also, pole testing and treating.

Painting steel towers is done by contract.

We have a local charitable organization which utilizes folks who can't find jobs because of various handicaps. We got them to bid on cleaning our glass meter covers. They gave us a price which is a fraction of our cost to do the same work, work which is not too thrilling for our own employees, trained as they are in sensitive meter testing.

We hire contractors to do our aerial patrols of transmission lines by helicopter. A company employee goes along as a passenger and makes his observations into a tape recorder for later transfer to a typed report.

A contractor performs our aerial surveys on new transmission line routes and on existing lines, when required.

Line surveys are contracted producing plan-profiles without utilizing any of our precious company engineering talent. The plan view is made from the aerial photographs, eliminating many hours of drafting time.

Plat surveys of lots to be used for substations, store rooms, pole yards and office sites are done by contractors.

We design our own stores facilities, pole yards and construction force facilities, but contract their construction.

Later, I'll be talking about handling trouble calls in connection with line breakdowns. An essential element of tackling this problem successfully is the use of a contract answering service.

Spread out as we are, it is impractical to perform vehicle repair at a central location, so it is done by local garages. Major repairs are kept to a minimum by timely scheduled minor maintainenance at the local filling station.

Every utility does much of the work I've listed by contract, but my division probably holds the record for doing a larger share of our work by contract than anyone else. We probably do 80 percent of our work by contract. This allows us to exercise a very wide range of control with a minimum engineering and supervisory force. Our exceptional success in cost control of all areas of activity bears out the wisdom of this course.

TROUBLE

Trouble. Nobody likes trouble. "Don't give me no trouble!" The comment indicates that trouble comes as an undesirable gift. The most loathsome part of my job is reviewing trouble reports that cross my desk. Each report I view as a failure on our part to live up to our bargain with our customers to give them uninterrupted service. Each one is an insult to my pride. They annoy me because they are not a gift. We produce most of these despicable things ourselves through our lack of proper, timely action and we inflict the tacky results upon our unfortunate customers.

One way many people avoid this problem is by just not making trouble reports. A transformer fuse blows. The customer calls the number listed in the telephone directory. If he's lucky someone answers. If he's a thoroughgoing child of fortune, a serviceman eventually appears, replaces the fuse, and he's back in business. The only record of all this activity is the serviceman's claim for overtime pay (which is never overlooked).

If we are to effectively treat with trouble, we must properly document it. To properly document it we must provide that serviceman with a piece of paper to write on. We should give him some indication as to what we want him to write on said piece of paper. We should indicate what that serviceman is to do with the piece of paper once he has finished writing on it.

We have tried several designs of trouble report forms over the almost 30 years we've been reporting trouble. All bad. Better than nothing, but none of them gave us the information we really needed to deal with the problem. Worse, they failed to clearly identify the problem.

So we sat down and thrashed it out, as they say. (Thrashing is always done in a sitting position.) What did we want the report to tell us?

Causes. We wanted it to identify the cause of the trouble.

Where? Locate the trouble in such a way that it can be pinpointed by an employee not intimately acquainted with the local area.

How much? How much line was out, how many customers out of service? How much equipment damage?

When should such a report be made? Since this report is used to cross-check overtime claimed, it must always be made whenever overtime is claimed, whether trouble is involved, or not. For instance, a trouble report is made whenever anyone claims overtime for planned work, or for transmission switching ordered by the system dispatcher.

Anytime anyone goes to the scene as a result of the lights going out or dimming, a trouble report is made, whether overtime is involved, or not.

Now we have a simple guide which may not always be followed but can never be misunderstood.

> Anyone
> Anytime
> Overtime
> Lights Out

Before plunging into an analysis of trouble, and since our report is dual-purpose in that it reveals both overtime and trouble, let's talk about overtime.

Overtime is bad. Most overtime is an unnecessary waste of money. Each hour of overtime is paid at a premium rate, accelerating the waste. Due to work rules, many overtime hours are paid for by the company in which no work at all is performed. Overtime promotes friction between the employee and the company and between the employee and other employees. Few employees feel that they get just exactly the right amount of overtime. The company shouldn't be satisfied with the overtime any single employee gets. And it ain't. Every employee feels that he gets too little, or too much overtime. He holds a similar opinion of every other employee with which he is acquainted. Wives of overtime-subject employees are equally opinionated.

Overtime is bad. The conscientious employee makes himself available for overtime service, gets a lot of it, and completes the year with more pay than the president of the company. Because he is so conscientious, you ask him to accept a promotion to service manager. He tells you to blow it out your homesick nose. You do. So you end up with some much less conscientious employee in the position of leadership, and management slides further into the Sea of Despond.

Overtime is bad. It is a sickeningly accurate measure of your failure to properly maintain your system and to prepare it to "weather" weather. It is a measure of management stupidity, indifference, incompetence and laziness.

How can we best manage this insolent measurement of our own rottenness, and our incapacity to manage?

Eliminate all overtime.

Having failed, manage the minimum amount remaining so that it is as equitably (not equally) divided among all employees subject to overtime, as possible.

Eliminating overtime requires eliminating the causes of overtime. I'll discuss that later on.

Managing the irreducible amount of overtime left after you have eliminated the eliminatable, is accomplished by the application of common sense (a very uncommon quality).

Bernard Baruch said that to make money in the stock market one must buy low and sell high. The way to equalize overtime is to assign most of it to those having the least of it.

Almost all overtime is charged by first- and second-class linemen and servicemen. Make a list of all these people and keep a tally of their overtime hours, year to date. Every month make a call-out priority list. It will essentially be a reverse of the first list. Low man goes to the top. Top man to the bottom. Now this must be tempered by your knowledge of each man's capabilities. You don't send a green newly-crowned, second-class lineman out at night by himself to handle an auto accident where energized conductors are wrapped around everything in sight. You don't send an old "honorary" first-class lineman with bad feet on such an assignment either. You don't send an after-five alcoholic out to do anything. But you can send the greeny and "foots" out to turn on a customer who has been turned off for nonpay and is now ready to pay up his bill *and* the cost of the overtime trip. (Why can they always find the money for that, but can't find it to pay the bill in the first place?)

The dispatcher is given all this information so he can use proper judgment in deciding who to call first, second, and so on.

The dispatcher. Who's that? In a big, rich metropolis, it is a company employee assigned to be on duty at all hours for just that purpose. Of course, more than one employee is required to cover all 24 hours.

But that's for rich folks. Out here in the country we can't afford such luxury. In a small one-serviceman town, the dispatcher is the local manager, or the serviceman, himself. If overtime gets out of line you analyze the causes. If tree trimming, grounding and preventive maintenance are indicated, you trim, ground and maintain. If it's due to a surplus of turn-ons after hours, you choose up sides with the local manager and hold a stabbing. If he persists, you assign him to take the surplus calls himself, until he gets tired of being uncooperative.

The same applies for two-servicemen towns.

In towns operated with three or more servicemen and where the construction force is headquartered (a district office town) it is a good plan to hire an answering service to handle after-hours calls.

We have had good luck with a small family-run answering service whose other clients are the local rural electric cooperative, the veterinarian, and a score or so of doctors. The local fire department is a possibility. The local rescue squad may come to *your* rescue.

If you've never tried this, you'll be full of doubts. "They aren't familiar with our system. They wouldn't understand enough about the way electricity operates to react properly in an emergency. They don't know the idiosyncrasies of our people." We have found quite the opposite. In a week or two the answering service is doing a better job than we ever did.

The answering service can screen calls better than we can. In a very professional, impersonal manner, they can divert nuisance calls to office hours which *we* would feel pressured into taking on overtime.

Whenever a major storm descends upon us, company supervisory personnel go to the office and call the answering service, relieving them of duty until the emergency is over. They do this by unflipping the switch which is flipped at 5:00 p.m. each night diverting all calls to the answering service. However, we have had storms in which it was 45 minutes before our supervisors could get to the office to flip the switch. The answering service performed admirably in the meantime.

The answering service maintains a log of the calls it places to our servicemen and linemen on which is noted every instance of "no answer," "not at home," "sick," or "refused to go out on the call." These are reviewed monthly and appropriate action taken, where indicated. We do not require our folks to "stand by," but we do require them to notify us (or the answering service) if they are going to be unavailable for an extended period during after-hours. If it becomes apparent that a man is going to be unavailable consistently, then we conclude that he really doesn't want to work for a utility company and be aggravated with helping us provide service to our customers on a 24-hour basis. And we tell him so.

The answering service is appreciated most by the wives of our servicemen. For years they suffered being kept awake all night by the ringing phone whenever it stormed. Now all calls are directed to the office number after hours and diverted to the answering service which places a single call to the serviceman or lineman to be dispatched on the trouble call. After that, the communication between serviceman and answering service is carried on over the company's two-way radio. The service is equipped with an extension of our base station.

Before instituting the answering service, two or three servicemen might show up at the same trouble spot, each called out by a different out-of-lights customer. This was not only expensive, it was unsafe. If three men answered the same call, there were probably two other trouble calls going begging while we untangled ourselves from our underwear.

We no longer have to pay home phone bills for numerous servicemen and linemen subject to a call-out.

The answering service permits more efficient utilization of manpower and equipment, since the service is in position to have all the facts available in one location.

Safety is enhanced with constant availability of contact with the dispatcher.

Our customers are better served. One number connects a customer in trouble with the company, just like calling the fire department, or the police. The lights come back on sooner, too.

In some cases local offices can be tied in with this system so that if the local company phone is unresponsive, the call is automatically diverted to the district office phone, thence to the answering service dispatcher.

Now let's get back to that trouble report form (see Fig. 1).

Some trouble calls are not controllable. They should be segregated from those which are controllable. We need to report them and categorize them, but they should not be allowed to cloud our view of those trouble calls which allow for corrective measures on our part.

So the report should be clearly divided with Uncontrollable incidents in one section and Controllable in another.

TROUBLE REPORT

DATE _____ NO._____

Geo. Code_____

LOCAL OFFICE_____

LOCATION_____ ZONE #_____

UNCONTROLLABLE

☐ 1. Pl. Work   ☐ 2. Disp. Req.   ☐ 3. Fire

☐ 4. Veh. Coll.   ☐ 5. Ctst. Wthr.   ☐ 6. Contrctr.

☐ 7. Van. Var. Aml.   ☐ 8. Tr. P.O. Serv.   ☐ 9. Trans. Sys.

☐ 10. Dist. Sub. Sta.

CONTROLLABLE

Primary

☐ 11. Lightning

☐ 12. Trees

☐ 13. Overload

☐ 14. Equip. Fail.

☐ 15. Bad Const.

Secondary

☐ 20. Trees

☐ 21. Bad Const.

Miscellaneous

☐ 24. Meter & Base

☐ 25. Cust. Accom.

Transformer

☐ 16. Lightning

☐ 17. Trees

☐ 18. Overload

    Amp_____kv_____

☐ 19. Bad Const.

Service

☐ 22. Trees

☐ 23. Bad Const.

☐ 26. St. Light.

☐ 27. Other (Describe Below)

Time Ser. Restored _____ Length Outage _____ min. No. Cust. Aff. _____

R.O. Subm't. ☐Yes ☐No Acc. Rpt. ☐Yes ☐No Imm. Inter. Rpt. ☐Yes ☐No

U–E–D–F–O–Report ☐ Yes ☐ No

*Transformer— includes riser, cutout, arrester, trsf. and leads.
Primary—all circuitry and devices of primary voltage, except transformers.
Secondary—from trans. to last supporting str.
Service—from last support str. to weatherhead.

Employee Name	RT	OT

Signed _____

Noted _____ Loc. Mgr.

and _____ Dist. Supt.

Apprd. _____ Dist. Mgr.

REMARKS: _____

Fig. 1

When we say Uncontrollable we don't mean that there is no control at all of these incidents, but that they do not comprise troubles resulting from inadequate action on our part which might have prevented the occurrence of those trouble calls. Fire, for instance. The police call and ask us to disconnect service to a building which is burning. We have no choice but to respond. But the event must be recorded. The fact that there was indeed a fire must be verified. We need to know that our facilities in no way contributed to causing or intensifying the fire. We need to record that service was discontinued to the structure involved, if it was, and when it will probably need to be reconnected.

We can't prevent drunks from driving their autos into our poles. These dreary events are classed as Uncontrollable. But we need to record them in order to collect damages from the driver. We need to evaluate whether the pole is poorly located, inviting repeated attacks.

When the system dispatcher calls and orders a serviceman to perform switching, or to check a breaker, we respond without question. We can't control that activity, but we must record it to verify the labor involved, whether regular time or overtime. Also, we need to keep informed as to why each request is made. It may involve system problems or changes which we need to take into account.

Planned work done on overtime must be approved before it is undertaken so when the overtime involved is reported, it is presumably justified and should be considered Uncontrollable. But it must be reported, nevertheless, and categorized.

Contractor damage is reported as Uncontrollable. We can't keep the peckerwoods from digging up our cables, digging out our poles or driving their bulldozers through our guy wires.

We have no control over faults which occur on the customers' wiring and are reflected on our system precipitating call-out of our people. So we have a category called "Trouble past the point of service." In such cases, we need to review our protective coordination and we need to be sure the customer is billed for our expense if that is in order.

Squirrels, snakes, birds, and hunters with lousy aim cause their share of grief. We report their activities under Uncontrollable. But each case is reviewed to be sure our clearances are proper and that we have done all it is practical to do to protect our facilities.

We do not consider that our local people are primarily responsible for avoiding trouble on transmission lines or in distribution substations. This responsibility rests largely at the division and general office level. So troubles on these facilities are reported as Uncontrollable, with a separate category for each.

Finally, we report as Uncontrollable trouble occurring as a result of catastrophic weather. Tornados, wide-spread sleet storms, Noah floods, blizzards, Chicago fires, locusts, pestilence and The End of the World are reported as Uncontrollable even though it is common knowledge that they are the direct result of voting Democrat. Do I mean that during a catastrophe we require the poor bedeviled workers to submit a trouble report for each individual case of trouble? Yes, we do. To the extent possible. We need this information in order to analyze how well our defenses held up to the elements and in what areas we need to lay more stress.

A catastrophe is declared and defined by the division office after it is past. We define a point on the clock as being the start and another point on the clock as its close. All catastrophe-related trouble reports within this time bracket are recorded under the category "Uncontrollable, catastrophe." No overtime hours charged by an individual during a catastrophe are recorded against his allowable overtime in our overtime control program.

During the past 12 months the Uncontrollable overtime amounted to 53.2 percent of the total overtime charged, including overtime charged to one catastrophe. If we eliminate the effect of the catastrophe, we had 33 percent Uncontrollable overtime.

Immediately, the thought comes to mind that if Uncontrollable were 67 percent, two-thirds of our trouble, instead of one-third, we would needs be about controlling Uncontrollable a little better. We would then turn lots of attention on each Uncontrollable category; first, to validate the accuracy of the reporting; and, second, to attack the causes and reduce their incidence to a level where we would be satisfied that the only cases of trouble left were truly Uncontrollable.

We have intensely analyzed our Uncontrollable overtime and are satisfied that we cannot expect to reduce its one-third share of the total significantly.

That leaves two-thirds that is fair game. Every item under the Controllable section is under constant attack. To make an effective attack, the categories must be reported in a way which permits us to easily, clearly, and quickly visualize and classify the sources of our problems. Here's the way we went at it.

We looked at the sections of the system. There's the generating plant. Forget it. Let someone else worry about that. We've already decided to reserve the transmission system and the distribution substation for the division and general offices to worry about. That leaves the primary, the transformer, the secondary, the service, and the ever-popular miscellaneous.

All controllable causes of trouble are covered by the following five-item list:

> Lightning
> Trees
> Overload
> Bad Construction
> Equipment Failure

All five apply to the primary. All but one of the five apply to the transformer. Only Trees and Bad Construction apply to the secondary and to the service.

We do not list equipment failure under transformer, because we don't think any of us is competent to judge that a burned-up transformer gave up the ghost purely as a result of deficient manufacture. When that does happen, the event is followed by fault current which destroys any evidence which might have given us a proper clue as to the incipient cause. If a new transformer fails within 10 days after it is energized, this is reported and we assume it is an equipment failure; otherwise we assume that the transformer failed as the result of a previous lightning stroke or overload, or that water seeped into the oil.

Now let's further define our geographic categories. Under Primary we include any event which produces damage to the primary circuit or which operates a protective device in the primary or at the substation.

Under Transformer we include any event which produces damage in the zone starting at the tap onto the primary, thence along the riser to the transformer, the transformer, thence along the secondary risers up to the taps onto the secondaries, or the service (if there are no secondaries). This zone includes the cutout, the lightning arrester and the transformer grounding system. If the transformer fuse blows and there is no reason to suspect that the cause was in the secondary or a service, the trouble is checked under the Transformer heading. If the fuse blows and the cause is determined to be on the secondary, the trouble is checked under Secondary. The same goes for the service.

Summarizing, if the cause of the trouble is on the primary, the trouble is checked under the Primary heading. If the cause is in the transformer zone, it is checked under Transformer. If on the secondary, check under Secondary. If on a service, check under Service.

Under Overload in the transformer section, we print AMP, _____, KV _____, so that we can learn the extent of the transformer overload. This leaves only Miscellaneous. Under this heading we have four items:

> Meter and Base
> Customer Accommodation

Street Light
Other

If the cause of the trouble is in the meter, which we own, or in the meter base, which we own, it is checked under this category. If the cause is in the entrance, which the customer owns, or in wiring past the meter base, we check under Customer Accommodation, or Other, as the case may be.

Customer Accommodation catches all the after-hours turn-ons. It also catches all those little old ladies who swear when they call in that everyone's lights are out all up and down the street. When the serviceman arrives he finds a circuit fuse blown and tells her to call an electrician. We'd like to change the fuse for her ourselves, but Seevers won't let us. He says this is a quagmire of bad public relations both with the affected customer and with the electricians we would be screwing. We get very, very few of these calls anymore.

The remainder of the form is for other pertinent information.

The date
The number assigned to the report
The local office involved
The location (usually just the street or road)
The zone

The zone. Early we found that we needed to identify our trouble with respect to locally-recognized "lines" or "sections." The "Jonesville Line" has a great deal of significance to the folks who work in the office that serves Jonesville. All the troubles on the Jonesville Line are considered as a whole. The Jonesville Road may be the favorite of teenagers trying out their new cars. Lots of auto accidents on the Jonesville Road. The old Jonesville Line may be 60 years old and prone to fall down. The trees didn't get trimmed on Jonesville Road last year. It's been three years since they had a haircut. Starting to have tree-caused outages on Jonesville Road. Better get out there and nip the worst until the trimmers get here next Spring. We were just ready to start grounding on Jonesville Road when the G.O. instituted their latest austerity program.

You can see the benefit in being able to segregate and identify all the troubles on Jonesville Road. So we set up zones out of each substation, dividing the lines out of that station according to the locally-accepted understanding of the segmentation of their system. Five zones per station is usually maximum.

If, after a year, it becomes apparent that Zone 3 is giving more than its share of troubles, we go to work analyzing all trouble reports marked Zone 3, and look for solutions. The analysis will always point in the direction of one or more of our defined trouble categories: Lightning, Trees, Overload, Equipment Failure, or Bad Construction.

Further defining the location of the trouble is the geographic code. Our mapping is based on U.S. Geodetic Survey quadrangles, scale 1"=2000'. Each quad is divided into 288 blocks, 24 vertical, 12 horizontal. Each block is placed on a 2' X 3' sheet at 1"=100' and covers a space in the center of the sheet 18¾" X 30¼", leaving room for duplicated lap-over around the edges. The 18¾" X 30¼" is defined by four hash marks. A clear plastic overlay divides this rectangle into ¼" blocks. The blocks are lettered A–Z, starting at the lower, left-hand hash mark and running vertically and horizontally until the 26 letters are used up, then repeat, using the entire alphabet three times vertically and five times horizontally with letters to spare. So we have fifteen 6½" squares on each 2' X 3' map. The squares are lettered A–O, left to right, starting at the top row. Now, any point on the 2' X 3' map can be located within 25' by a three-letter code. First letter identifies the 6½" block. The second letter identifies the vertical position on that block and the third letter identifies the horizontal position.

The quad is 24 maps high, 12 wide. So, we use a bi-quad, two quads side-by-side, giving 24 maps high, 24 wide, and assign letters A–X vertically and horizontally. Now, two letters identify the 1"=100' map in question on the bi-quad.

A grid of bi-quads 26 high, 26 wide, covers most of my company. This grid can be chosen to fit the area involved. Then two letters will identify the bi-quad in question.

Seven letters identifies any point within 25 feet. Each 25' square can be divided into 1 foot blocks and lettered A–Y to identify any point within one foot. This would be useful in connection with computer mapping. It would add an eighth and a ninth letter to the code. Twenty-five feet is close enough for our purposes at this time.

Other schemes would do just as well, but *some* scheme must be used to identify the location of trouble.

This seven letter code can then be shown on all the customer's computer data, including his service bill. The code can be mailed to him on a stick-on in his bill to be attached to his telephone. Then, in case of trouble if the dispatcher has any trouble locating the customer geographically (as we often do with customers who have just moved into the state, for instance), he can ask the customer to read the geographic code stuck onto his telephone. This code enables servicemen to pinpoint the location of turn-ons, turn-offs, and meter change-outs.

Besides the benefits already listed which the code offers, it enables the division office to pinpoint trouble locations within zones which may be under close investigation.

The serviceman may carry with him a roll of 1"=100' maps covering the entire zone he is working on a particular day. But that is really not necessary. A maximum of five 1"=2000' maps will cover any zone and since these maps are marked with a grid showing where the 1"=100' maps fit, the serviceman can usually orient himself well enough by referring to his 1"=2000' maps. If that is not enough, he can communicate by radio with the dispatcher, who has all of the 1"=100' maps handy and can give him detailed directions whether in connection with normal daytime activities or after-hours trouble calls.

In another section of the report we should indicate:

Time service is restored

Length of outage in minutes
Number of customers off
Did this trouble generate the following reports?
    Routine Order
    Accident Report
    Immediate Interruption Report
    Underground Failure Report

With this information we can cross-check the other reports.

In another section of the report we must have the name(s) of the individual or individuals who responded to the trouble call, how many regular time hours were charged and how many overtime hours were charged. The report should be signed and then approved by the appropriate supervisors before being sent in to the division office.

Finally, a generous space should be allowed for remarks. The serviceman's remarks can be invaluable in producing a solution to a sticky trouble problem.

With this report, the division receives adequate information to analyze all trouble in detail. However, the district is required to compile all trouble reports once a month into a monthly district analysis report (see Figure 2). Each category is given a number, from one to twenty-seven. The categories are listed down the left-hand side of an 8½" X 11" page, preceded by the corresponding code numbers. The next column lists the hours, regular or overtime, spent during the month on each category. The next column lists the hours spent during the preceding 12 months. The next column lists the hours spent this month one year ago. The final column is the first column, plus the second, minus the third. That is the hours spent on each category during the past 12 months.

The same form is used at the division level with six additional columns. The first column is the sum of column four from all the districts, divided by the total number of customers in the division as of 12-31 the previous year. This is the hours per customer spent on that category of trouble during the past

DIVISION MONTHLY ANALYSIS MO. August YR. 1980

No. Reports This Mo. 360 TROUBLE REPORT – MAN-HOURS Customers 87659

| | HRS. THIS MONTH | | | | MAN-HOURS PER 1000 CUSTOMERS | | | | | |
| | This Month | Prev. 12 mos. | Mo.Yr. Ago | 12 Mos. To Date | Div. | 12 months to date | | | | |
						1	2	3	4	5
UNCONTROLLABLE										
1 Planned Work	6	276	14	268	3.0			3.9	5.2	5.7
2 Disp. Requests	10	54	1	63	.7		.3	.4	1.7	.7
3 Fire	52	315	10	357	4.1	4.1	2.2	4.6	4.9	3.5
4 Vehicular Collision	72	1059	29	1102	12.6	12.5	7.2	14.1	11.3	16.5
5 Catastrophic Weath.	19	1753		1772	20.2	25.8	7.9	27.2	16.7	21.6
6 Contractor	55	395	7	443	5.1	4.1	3.6	5.5	5.3	6.2
7 Varmint, Animal, Bird, Vandalism	37	344	20	361	4.1	3.4	3.7	2.6	3.3	7.6
8 Trouble past point of service	15	145	10	150	1.7	.7	1.2	6.1	.5	1.9
9 Trans. System	19	31		50	.6	.8		.3	.5	1.0
10 Dist. Sub.	5	89	1	93	1.1	1.0	.6	2.2	1.0	.6
TOTAL	290	4461	92	4659	53.2	52.4	26.7	66.9	50.4	65.3

PRIMARY										
11 Lightning	124	768	234	658	7.5	8.4	6.7	4.6	6.9	9.3
12 Trees	99	435	77	661	7.5	6.8	3.1	10.4	8.6	7.8
13 Overload	1	157	36	122	1.4	2.2	1.9	1.2	1.2	.1
14 Equip. Failure	9	421	28	402	4.6	3.5	3.1	2.1	5.5	7.3
15 Bad Const.		82	6	76	.9	1.7	.2	1.0	.9	.3
TRANSFORMER										
16 Lightning	146	768	191	723	8.2	4.1	5.0	8.0	12.4	9.6
17 Trees	16	159	12	163	1.9	.1	.3	.9	3.8	3.3
18 Overload	168	451	70	549	6.3	4.6	5.5	3.8	9.9	4.7
19 Bad Const.	5	67	7	65	.7	.3		1.8	.5	1.3
SECONDARY										
20 Trees	32	206	30	208	2.4	2.3	1.3	1.3	4.1	1.6
21 Bad Const.	23	55	1	77	.9	.6	1.4	.7	1.0	.5
SERVICE										
22 Trees	38	344	56	326	3.7	3.7	2.1	2.6	5.0	3.7
23 Bad Const.	18	108	14	112	1.3	1.3	.4	.4	2.2	1.3
24 Meter or Base	10	91	7	64	1.1	.7	.6	.8	1.6	1.1
25 Customer Accommod.	53	491	24	520	5.9	5.3	5.7	.4	8.9	5.3
26 St. Lighting	4	18	7	15	.2	.1	.5	.5		.2
27 Other (Describe)	40	717	56	701	8.0	11.4	3.6	10.9	8.3	4.6
TOTAL	786	5542	856	5481	62.5	57.1	41.2	50.8	80.8	62.0

Fig. 2

12 months. The remaining columns are the result of similar calculations for each district.

The first 10 lines comprise the Uncontrollable trouble and the sum of each column appears just beneath the upper section of the report. The lower section of the report comprises the 17 Controllable Trouble categories and the sum of each column appears across the bottom of the page.

Now, we have this octopus by the tentacles. The figures on our division report permit us to make all sorts of comparisons. For instance, I have already pointed out that we have 33 percent Uncontrollable and 67 percent Controllable.

The next thing is to look at the Controllable. We find which district is high, which is low. Next we look up the page to find out which category, or categories, make the high district look so bad. I find that the item most out of line with the division as a whole is item 16, Transformer—Lightning. This district has not completed its grounding program. One other district is also high in this category, even though their total controllable is right on the division average. I know that this district is farther behind on grounding than the first district.

The districts that have completed their grounding, what about them? Well, one is right on the division average, and the other two are about one-half the division average. We check and find out that the average district completed its grounding but, due to soil conditions, had to leave ground readings much higher than the other two districts. We may have to go back and lick our calf over again.

One district's total is only two-thirds of the division average. We look up the page and find that the outstanding variations below division average appear in the categories: Primary—Lightning, Trees; Transformer—Lightning, Trees, Bad Construction; Secondary—Trees; Service—Trees, Bad Construction; and finally, this district lists less than half as much time per customer under Miscellaneous as the division average.

Let's look at this thing from another angle. Let's take the

category Trees and see how the districts compare. Under Primary—Trees, three districts about average, one district 50 percent over average and one district only 50 percent of average. We need to consider reapportioning our contractor trimming budget.

Under Secondary—Trees we find only 25 percent as many hours as under Primary. Considering the fact that we don't intentionally spend any contractor dollars on secondary trimming, and that outage due to trees in secondaries is limited in scope, that ain't too bad.

Bad Construction accounts for only a very small part of the total and is highest on services. Here's where bad connections and splices show up. The district with the best preventive maintenance record has only one-third as much trouble per thousand customers as the division as a whole, and only one-fifth as much as the district with the worst preventive maintenance record.

Overload is relatively small on Primary, but significant on Transformers. One district has almost three times as many transformer overloads as the district with the best record. The other three are about equal. We'll have to go after that high district and get that figure down. We can probably learn something from the district with the low incidence of Overload (again, this is the district with the best P.M. record). One district doesn't seem to be bothered with primary overloads at all. I wonder why? Two districts are noticeably high. We'll have to analyze their individual trouble reports and come up with an answer.

Everyone seems to suffer equally from service-related troubles.

Customer Accommodation is about equal for three districts at 5 man-hours per 1000 customers. Why do you suppose that the fourth district has only .4 and the fifth 8.9? We find that the fourth district is less active in customer turnover and has a tighter policy of resisting after-hours turn-ons. The fifth district is very active and has slipped into the bad habit of almost jumping in response to dead-beats' demands to be turned on after being turned off during the day for nonpayment.

Let's let the report tell us where the bulk of our Controllable trouble occurs.

Primary—35%
Transformer—27%
Secondary—5%
Service—10%
Customer Accommodation—9%
Street Lighting—0.3%
Other—13%

From this we see that we need to concentrate on improvements to the Primary and to our Transformers.

Looking at the source of our problems on Primaries and Transformers, we see:

Lightning—40%
Trees—24%
Overload—20%
Equipment Failure—12%
Bad Construction—4%

Bad Construction is not our problem. Equipment Failure is not something we can expect our field forces to solve. The bulk of our overloads are on transformers. We need to develop ways to avoid having air-conditioning loads slip up on us. Tree problems seem to be under pretty good control. Just keep the program equalized among the districts. That leaves lightning as enemy number one. We must do a better job of effective grounding.

Let's analyze Uncontrollable. Planned work (on overtime) varies wildly among the districts. Since none is undertaken without my personal approval, it is perfect.

Switching requests from the system dispatcher are not within our control.

Note that fire seems to play no favorites. But vehicular collisions vary more than two to one.

We see that our massive storm hit some districts harder than others.

Contractors, like fire, plague each district about equally.

So do varmints. Maybe we should have included contractors in the varmint category.

Trouble past the point of service. One district seems to be inclined to do more "public relations" troubleshooting than the others. Better not let that get out of hand.

The transmission system must be in good shape. So are our distribution substations.

You can see that our report gives us a wealth of information on our performance in giving our customers what they pay for—quality service.

When a particular item is out of line, we backtrack through the district reports and the initial trouble reports to determine where action is needed and what improvements are indicated. We may find that trees are out of line in a particular district. By backtracking we may isolate most of the trouble to one zone. Instead of retrimming the whole district we attack that one zone, at far less cost.

If we find one district high on transformer lightning, we backtrack. We will probably find that grounding is incomplete. But even where it is complete, we will find that in certain areas of the district we just couldn't get our grounds down to acceptable levels. We got them down from 100 ohms to 30 ohms. That's an improvement. But now we need to decide how to improve on those 30s. But just in the one area. Not over the entire district. More money saved. And what we spend will be spent effectively.

Like the man who lies awake nights thinking how he used to suffer from insomnia, I don't know how we managed before we instituted this trouble control program.

THE SERVICE CENTER

The Service Center. Now, doesn't that sound better than "The Storeroom," "The Barn," "The Store Yard," or any other such phrase left over from the days of horse-drawn line trucks?

The service center is where we concentrate the means to provide quality electric service to our customers. Service is the key word. The center is not to be designed for our comfort or convenience. Everything about the design should enhance service to our customers. I once heard a company official state to a group of employees that the purpose of our company was to provide a pleasant place for our employees to work and make a good living. Always anxious to ingratiate myself with management, I replied that he was crazy as hell if he thought that. "The purpose," I said, "is to make an adequate profit for our owners. The only way to do that is to provide service for which our customers will willingly pay. To provide that service requires competent employees. To get them, we must provide a pleasant place to work and insure them a good living."

Good service to our customers is not enhanced by air-conditioning stock crossarms or bolts. A transformer which will spend its working life hanging on a pole in hot weather, cold weather, rain and snow, does not need to be kept in comfort while awaiting its cue to go on stage. Yet it has been common practice to house such materials indoors, out of the weather, heating them in winter, and in some cases even cooling them in summer. Why?

Same old reason. "That's the way it has always been done." Also, it is more convenient to pick up a nice clean crossarm and load it on the truck than to dig it out from under a pile of snow.

The point is that in designing a new service center we should spend our available money to the best advantage. We cannot afford to waste money on facilities whose only justification is that we have always provided such facilities in the past. We should, instead, look at each item which will occupy the new center and determine the most economical, efficient, practical provision for that item.

The "items" to be considered are materials, workers, and vehicles. Let's provide for the workers first.

Adequate washroom facilities for the number of workers involved is essential. Toilet, urinal, and wash basin. A shower might be justified in some situations, but we haven't come across such a situation yet.

Office facilities for the superintendent and the stores clerk. Each in his own office. No one should be expected to perform through another's telephone conversation. Besides, the superintendent should have an office where he can chew people out in privacy.

Quarters for the construction force. Here we should provide an adequate writing space for the foremen to complete necessary reports. Time, material and efficiency. We provide an 18-inch deep counter down one side of the assembly room. A drawer is mounted under the counter every four feet. The foremen use folding chairs which can be moved out of the way when the counter is not in use.

The assembly room must be sized to accommodate the number of workers involved; 15-20 square feet per person is sufficient. In this room are the writing counter, lockers, and a table about 3' X 6'. There should be only enough folding chairs to accommodate the workers and 4 or 5 guests.

A closet. Got to have a locked closet to hold tools, spare rubber goods, and small "disappearing" items. I tend to steal one roll of electrical tape each year, so it's a good idea to keep it locked up where I have to work at it to replenish my supply. Don't make it too easy for me.

That's all. An assembly room, two offices, a washroom and a closet. Sounds Spartan. It's meant to be. No pool tables. No prayer chapels. No saunas. No card tables. This is a place of business. Not an Elk's Club. The workers, if properly supervised, will spend no more than 5-10 minutes a day in the office area of the service center. It would not be in the best interest of our customers to equip it like the Old Mason's Home.

These four rooms and closet are heated and air-conditioned. No other air-conditioning is required. The only other heat required is for the shop area and the scales.

The shop area is adjacent to the office area but must be taller to accommodate the tallest vehicle. It should be limited in floor space to the very minimum necessary to work on one vehicle at a time and to maintain reclosers and transformers. The tall walls are ideal for storing hot sticks. (Remember hot sticks?) The only way to keep the shop from becoming a junk room is to size it just a little smaller than it really ought to be. By keeping it small we also limit the heat required. This area should be heated to 50 degrees. No more. It will be used by workers dressed for the outside weather.

Our vehicles do not require housing in heated, cooled buildings. They are operated in the open in whatever weather is at hand. They do need to be sheltered at night so that the workers will not have to load a wet, or snow-covered vehicle in the morning. Providing a shelter also keeps them out of the winter winds, making for easier starts. The shelter also keeps the vehicles out of sight of the public. A yard full of big trucks does tend to look trashy and invites theft and vandalism.

So we must build a shelter for our work vehicles. What kind of shelter? How big?

It should be sturdy to last a long time. It should be as inexpensive as possible. It should be ample in size to facilitate parking the vehicles and maneuvering them in the building.

This suggested to me a wooden frame with galvanized sheet steel covering. A long building with a two-lane concrete drive down the middle of it from one end to the other with an open doorway at each end—no expensive overhead doors.

The length is determined by the number of vehicles to be housed. Six trucks can be housed with room to spare in a barn 200 feet long. Any vehicle can be pulled out of line and driven past the others since we are using a two-lane driveway.

Now we have a structure 200 feet long with a concrete driveway about 22 feet wide running from one end to the other. The ends are open allowing free flow of air. This prevents damage to the structure from tornado-induced vacuum conditions. The height is sufficient to accommodate the tallest vehicle.

Add 8 feet to the width of the building down one side of the driveway and 12 feet to the width down the other side and you have storage space for an immense amount of material. The 12-foot side is convenient for long items, such as ground rods and anchor rods, and the 8-foot side can be used for materials requiring shorter space. An inclined, galvanized-steel flat roof covers the driveway and the 8-foot storage. The 12-foot section is covered by another inclined, galvanized-steel flat roof draining in the opposite direction and only 10 feet or so above ground level. This produces a flat face starting at the line where the roof over the 12-foot section joins the building up to the peak of the main roof. This is covered, not by galvanized steel, but by transluscent panels which allow light into the building throughout its entire length.

No shelving is required. The material is stored in the boxes it is shipped in and stacked with the one opened box on top of the stack. The 10-foot bays of the supporting structure provide convenient separation for the various types of materials. Each 10-foot bay becomes a booth in which is set up a "shop" for a particular type of material. For instance, there is a shop for grounding materials, a shop for anchors and guying, a shop for arresters, for cutouts, etc.

As each truck is loaded in the morning it moves along the driveway in the "passing lane" from shop to shop picking up the desired items. This makes for quick loading and facilitates accurate material records. At the end of the day the process is reversed.

All boxed materials are stored in the barn, out of the weather. This leaves only two items, transformers and conductors, which we haven't provided for. They are stored in the open on concrete docks. The edge of each dock from which materials and equipment are loaded and unloaded is 42 inches high to match the height of the beds of the trucks which back up to them. You can have as many docks as you want. They are 10 feet deep if single-side loading is contemplated, and 20 feet deep if loading is to be from both sides. We find that two singles and one double is sufficient for the 200-foot barn complex. All docks are about 200 feet long. Rocked drives 55 feet wide are provided between docks.

No matter how much space you provide in the barn and on the docks it will all be filled soon after construction is complete. You want your people to have adequate space to store all the materials they will need. But the only way to control your stock is to limit the space available to store it. Therefore, it is important to calculate the proper space requirements before construction to be sure you have enough without providing space for overstocking.

The conductor scales are located in a protective shed and must be provided with a minimal amount of heat. Antifreeze is used in the drain trap. The "scale house" can be located anywhere your people want it, but we like to have it adjacent to the end of one of the docks closest to the service center entrance, where all trucks must pass by it when entering or leaving. This reminds the driver to "weigh in" and "weigh out." The scales can be mounted at ground level or at dock level. Whichever you pick, somebody will say it should have been mounted the other way.

The entire complex must be surrounded by a chain-link fence. The gate should be motor-driven and operated by a company key. It should also be possible for the storeskeeper to operate the gate by remote control in response to a radio message from the gate requesting entry.

You now have a basic description of each of the elements which go to make up an efficient, inexpensive service center, serving a district area in the 30,000 customer class. The elements can be expanded to adapt it to larger service areas. But it is my observation that much larger service areas would be better served by breaking them up into smaller service areas with a service center for each. I believe a service center no larger than the one I have described can service up to 50,000 customers. The number of employees working out of a single center should be no more than 20. Any more will produce traffic jams and confusion. The area serviced from such a center may involve travel of only a few miles from the center, if located in a congested population center. If located in an area involving small cities or towns, and lots of rural space, the maximum travel from the center may be as much as 50 miles, an hour and a half.

The location of each of the elements of the center depends on the size and terrain of the available lot. It can be located on a lot as small as three acres. Maximum required is four acres. If located on a slanting lot, the docks can be built in a terraced fashion with the driveways between them sloped gently to provide drainage.

Be sure that the roofing nails are installed with neoprene washers or the roof will leak.

It is a waste of money to paint these galvanized-steel buildings. The paint will fade in a year or so and you won't be able to tell it was ever painted.

A large plastic lettered sign should be installed along the side of the barn facing the roadway. " _____ Utilities Service Center." Red letters about 18 inches high. Aim floodlights at the sign from ground level and switch them on and off with a photocontrol.

Not counting land cost, $200,000 will do it. That's a fraction of what it costs for "conventional" facilities.

But that's just the start of your savings. You'll save at least a half man-hour per man per day in loading and unloading time. Inventory is a snap with all that material spread out over hundreds of feet. And with boxed materials stored in the boxes you just count boxes and multiply by units per box and add the loose units in the top box.

Transformers are spread out and grouped by voltage and size. Conductor is spread out and grouped by type and size.

Actually, if the storekeeper does his job, inventory involves only counting odd items. He ought to keep a running inventory on all full-box, full-reel, etc. items.

You never have to move any item to count another item. No items behind or under others.

With this set-up our construction units are on the road by 8:15 to 8:30 in the morning and return to the center no earlier than 4:30 in the afternoon.

We still turn up discrepancies at stock inventory time. Our G.O. still screws up the computer records. But now our storekeeper has confidence in his ability to point out their errors and get them to make the necessary corrections. Oh, we make a tiny error ourselves now and then!

Finally, the service center design *looks* good. It gives the impression of order. It tells the public that we are in control of our materials. It bespeaks efficiency to the passer-by. It causes our workers to want to maintain an ordered appearance in the materials they handle.

What more do you want?

THE POLE YARD

We have given up on the railroads to deliver our standard length poles. Poles from 30 feet in length to 60 feet in length come by truck. We have given up on trying to get the pole companies to deliver in trucks with automatic unloading capabilities.

The railroad is out. That just frees us to locate our pole yards where *we* want them instead of trying to find inconvenient and costly locations near rail.

We can go out in the country and buy adequate space relatively cheap located near major highway facilities and central to the work area. 300 X 300 feet is sufficient.

Again we spread the material out to facilitate loading, unloading, and inventory.

We build a concrete driveway between two rows of pole racks. The racks are truck-bed high along the drive and slope down to the rear where we install an "I"–beam stop to stop the poles as they roll off the truck. There are three racks down one side of the drive; one for 30s, one for 35s, one for 40s. There are two racks down the other side; one for 45s, one for 50s.

The racks are made by installing concrete supports at the front and rear with intermediate supports as necessary, depending on the pole weights to be stored. Poles are then laid on these supports with the butt ends toward the drive to absorb the shock of poles being dumped off the truck. Four poles per rack for 30s, 35s, and 40s. Five poles per rack for 45s and 50s.

The driveway is about 14 feet wide with the center higher than the edges so that either side the truck drives down to unload poles the truck will be tilted toward the pole racks about

5 degrees. A heavy angle iron is mounted to the rack pole butts along the driveway end of the pole racks for the truck to snuggle up to. Just release the front and rear of the pole load simultaneously with the stays on the pole rack side already down, and the load slides off onto the pole rack as slick as a ribbon!

For goodness sake don't let them load the poles on the truck with all the butts pointed in the same direction! If they are interspersed, the whole load will come off straight and roll down the rack straight.

At the far end of the concrete driveway we go to crushed rock, and carefully lay out a roadway which allows a tractor trailer to exit from between the pole racks, clear them, and turn and circle back to the pole yard entrance. Along this route is plenty of room to store 55s and 60s on crushed rock, no racks. You don't need many of them.

Poles longer than 60 feet are stored at the last site rented for poles used in building the most recent transmission line. Don't mess up your distribution pole yard with those bastard sizes.

Since I set all my poles by contract, I don't have to concern myself with locating my pole supply close to my stores center. In fact, I don't want my contractor showing up at the stores center while my company forces are loading or unloading. They can load up on anchors, anchor rods, ground wire and moulding once a week, or so, and that can be at mid-day.

Like my service center, my pole yard is inexpensive, it looks good, it bespeaks order and efficiency.

The contractor loads poles onto his truck from the rear of the pole racks so that the poles continue to work to the rear, just like a cigarette machine.

Inventory time is a delight with only one size pole on each rack. Just walk down one side of a rack, count butts. Walk down the other side, count butts. Add the butts. That's all.

Cutest pole yard you ever saw!

EFFICIENCY RATING SYSTEM (E/R)

"Effective operation as measured by a comparison of production with cost in energy, time, money, etc."

So says Webster in defining efficiency.

So efficiency is a *measure*.

It measures *effectiveness*.

Effectiveness is *desirable results*.

How much *cost* to get results?

So far, so good.

But _____

How do we define and measure *costs*? *Results*?

The first thing that hits us is that results are *good*.

Costs are *bad*.

Optimize results.

Minimize costs.

Wasn't that easy?

The logic is easy to grasp, but putting this logic into practice is not so easy. It is, however, much easier than most managers think it is.

To establish the machinery to measure efficiency is a snap. Many methods have been devised and are already available. We will discuss later only the best method—mine.

After the system of measurement is successfully in operation and it tells you that your operation is lousy, generally, and miserable in particular areas, the whole process can stop, producing no more inconvenience to anyone. You may rest assured that "I told you so!"

"Sorry, sir, our tests indicate that you have a rapidly developing tumor which may be malignant."

117

"Good Lord, doctor, what are you going to do about it?"

"Do? Oh, we really hadn't given that any thought. Any treatment we might attempt would involve unpleasantness to you and your family. Let's just leave well enough alone."

"Dammit, doc, you've got to do *something!* I don't want to die."

"Well, why don't we treat a few of the more flagrant symptoms? Since you don't wish to die, we'll perform just enough treatment to keep you from dying. At least, not for some time."

"Forget it. I'm going to get a new doctor, fast!"

There you have it. The analogy puts no strain on the imagination as it applies to utility operation today.

"Insured a profit."

"Rest in peace."

The idea that the utility will always be insured a profit by a benevolent governmental commission made up of members who hope to run for governor is becoming our death knell. We simply go to them and explain that our expenses exceed our income. Whamo! They give us a rate increase.

This system directs our attention and energies toward demonstrating how high our expenses are. If, God forbid, our expenses were much less, the commission would direct a rate *reduction.*

The system optimizes costs, belittles results.

Ah, but times are changing. No longer is rate relief automatic. (Note in the term "relief" the tendency to become a body at rest.) You can't run for governor on a platform consisting of your generosity to the utility company at the expense of the voters. Rising utility rates are "news."

After losing the football game the coach announces at practice Monday afternoon that "We'll get back to the basics." Like it or not, utilities are getting back to the basics. You can't up income. You must reduce costs.

How? With the plausible excuse that unless we improve efficiency none of us will have a job, we now ask our people to *do* their jobs.

By this I do *not* mean:
>Hurry.
>Sweat more.
>Short cut safety.
>Mistreat people.
>Invoke fear.
>Cut pay, fringes.
>Mickey Mouse.

I *do* mean:
>Invite supervisors to join the game.
>Give the people the facts.
>Get the facts.
>Establish a plan.
>Work smarter.
>Show that everyone's time is important.

So much for philosophy. We've established that efficiency is desirable, essential to the well-being of us all, from the janitor to the president. Let's do it!

How?

Back to basics. Efficiency measures results.
>Identify the desirable results.
>Develop a measurement for each.
>Institute the program.
>Evaluate the data it gives you.
>Implement indicated improvements.
>Monitor the effects.

Now, one at a time.

Identify the desirable results.

The results we desire are work units. A meter reader's work units consist of meters read. A company president's work units consists of a variety of ever–changing tasks. We choose to confine our study to jobs consisting of *repetitive* work units, easily measured, recorded and *verified*.

In an electric utility, this includes the construction force, the servicemen, and the meter readers. Their work units largely dictate the work units of the rest of us—office clerks, engineers, superintendents, managers. Accurate evaluation of the performance of the first three groups will produce data allowing for evaluation of others.

Take meter readers. Their work can be measured by two units. RMW and RMR. Read meter-walking, and read meter-riding.

If he walks to the meter location from the previous meter location, he gets one RMW unit for each dial register read and recorded. If he rides to the meter location from the previous meter location, he gets one RMR unit for each dial register read and recorded.

Next, servicemen. Their repetitive tasks include eight units aside from meter reading. They are:

Turn off or turn on. No meter is hauled to or from the site. He gets one TON for a turn on, only. He gets one TON unit for a turn off, only. A "read and leave on" equals one turn on and one turn off, only. He gets *no* unit for "check meter reading generated by computer."

Meter change out. Meter is always hauled to and/or from the site. He gets one MCO unit for the following tasks:

 Set a meter and turn on.

 Add a meter.

 Remove a meter.

 Exchange a meter.

 Turn off and remove.

Collect. He gets one CLT unit for a turn off for nonpay. He gets one CLT unit for a turn back on after payment is made. He does not normally accept payment at the site. He gets no unit for making a telephone call to collect, or for making a visit to the site if he does not execute a turn off or a turn on at that time.

This collection policy eliminates lots of wasted effort, duplication of visits, and invitation to muggers.

Periodic substation inspections. He gets one SUB unit for each monthly distribution and/or transmission substation or transmission switching station check. He gets one SUB unit for each weekly transmission substation or switching station check. Only routine checks are credited.

Customer complaints. He gets one CMP unit for handling a high bill complaint. He gets one CMP unit for handling either a high- or low-voltage complaint. One unit may involve more than one trip to the site.

Relamp street light or customer outdoor light. He gets one LMP unit for relamping a COL. He gets one LMP unit for re-lamping a street light. He gets one LMP unit for replacing a street light relay or photocontrol.

Routine Order. He gets one RO unit for each routine order prepared, including easement, staking, and spotting entrance.

Monthly recloser or capacitor installation inspection. He gets one CKR unit for each monthly inspection of a single phase recloser or capacitor installation. He gets one CKR unit for each monthly inspection of a three-phase recloser or capacitor installation. Three single-phase OCRs at one site equals three units. A three-phase OCR equals one unit.

Construction work is covered by 23 units. Whereas each meter reader's units and each serviceman's units are tallied for each individual worker, construction units are tallied for each work force consisting of ten to twelve workmen, including foremen who work as a unit out of a single service center under the direction of a district superintendent. The workers are assigned jobs each day with the optimum number of workers and equipment for that job, or jobs. This varies from day to day. It is good to avoid set crews. Units may be tallied on a temporary basis for one month on a two- or three-man crew out of the work force which is allowed to stay together for a whole month for the purpose of evaluating that portion of the work force on a separate basis.

Construction units are recorded as I—Install, R—Remove, and T—Transfer.

Pole. One POL unit for installing or removing any height pole. The transfer unit is used to record straightening an existing pole, filling woodpecker holes, topping, etc.

Crossarm. One ARM unit for installing, removing or transferring a crossarm or MIF bracket.

Pole top pin. One PTP unit for installing, removing or transferring a pole top pin, neutral spool bolt, rack (any size), four-way pole band, or a one- or two-bell D.E. assembly. One PTP unit for adding a bell.

Anchor. One ANC unit for installing or removing an anchor, any type. No transfer unit.

Guy. One GUY unit for installing, removing or transferring a down guy or span guy, any size.

Cluster mount. One CLS unit for installing, removing or transferring a cluster mount. Four units for a two-pole steel platform.

Pin-type insulator. One PTI unit for installing, removing or transferring a pin-type insulator. Includes wrap lock, tape, tie, armor rods, etc.

Dead end. One DE unit for installing, removing, or transferring any type dead end, primary or secondary. Like treatment is given the following items under the DE umbrella:

> URD terminator, outdoor or indoor
> Pistol grip terminator
> T–Tap
> Underground splice
> Underground secondary terminal
> (Do not report with the underground service.)
> Underground neutral or concentric neutral terminal

Ground. One GRD unit for installing a ground initially or subsequently improving a ground. Report one unit per grounding installation, no matter how many rods are driven. This unit includes meggaring. One unit for removing a ground. No unit for transfer.

Lightning Arrester. One LA unit for installing, removing or transferring a lightning arrester or cutout. Credit one LA/R (remove) for each OCR, sectionalizer or oil switch removed for servicing, whether serviced at the site or returned to the storeroom. Credit one LA/I when unit is reinstalled.

Service. One 2/0S unit for installing, removing or transferring any service, overhead or underground 2/0 or smaller. (Don't report D.E.s or terminations—they are included in this unit.)

Service. One 4/0S unit for installing, removing or transferring any service, overhead or underground 4/0 or larger. (Don't report D.E.s or terminations—they are included in this unit.)

Street light. One COL unit for installing, removing or transferring any size street light or customer outdoor light. Report 2 units for each metal street light standard.

Guy wire protector. One GWP unit for installing or removing any type of guy wire protector. No transfer unit.

Transformer. One 50 unit for installing, removing or transferring any overhead or padmount transformer, 50 KVA or smaller, or any CT or Pt. Includes wiring.

Transformer. One 75 unit for installing, removing or transferring any overhead or padmount transformer, 75 KVA or larger. Includes wiring.

Oil circuit recloser. One OCR unit for installing, removing or transferring a recloser, sectionalizer, or oil switch for each phase involved. To illustrate, report 3 units for one three-phase OCR.

Capacitor. One CAP unit for installing, removing or transferring each phase of any size capacitor installation. To illustrate, report 3 units for a three-phase capacitor. For replacing one failed can, report one unit (I) and one unit (R).

Disconnect. One DISC unit for installing, removing or transferring each phase, any size disconnect or switch. To illustrate, report 3 units for a gang-operated air-break switch.

Conductor. Report one 2/0 unit for installing or removing each 100 feet of overhead conductor or underground cable, primary or secondary, any size, 2/0 or smaller. (Three conductor cable scores three times as much as a single conductor cable.) Report 1 unit 2/0 (T) per attachment on transfers. (For example, move phase from PTP to pin position, report 1 unit.)

Conductor. Report one 4/0 unit for installing or removing each 100 feet of overhead conductor or underground cable, primary or secondary, any size, 4/0 through 500 MCM. Otherwise, report same as 2/0.

Conductor. One unit, 750 MCM, or larger, etc. Otherwise, same as 2/0.

Cable. One unit CBL for each 100 feet of duplex, triplex, or quadruplex installed as secondary, any size. Do not report D.E.s separately—they are included in the unit loading.

There you have it. Two simple task units for meter readers, eight for servicemen, 23 for the construction force, 33 in all. Add to this the actual time spent by servicemen on preventive maintenance and you have the measure of 99 percent of the productive time of these gentlemen.

What about time spent in meeting? Travel time? Coffee breaks? Etc., etc., etc.

O.K. Let's discuss just what portion of a worker's time should be used in our measuring process. Obviously, it behooves us to eliminate all of his hours over which management and his supervisors have no control. Total available hours equals the total hours reported for payroll less vacation, holiday, sick leave, time worked on major storm trouble, whether regular time or overtime (overtime hours spent on regular work, such as working an extra hour to complete a job which otherwise would require a return to the site the following day, is included in available hours, as are the units involved), military duty, jury duty and time off to attend grannie's funeral. Hours spent on unusual work not involving standard units is not included. This includes temporary assignment to work on substation construction, transmission patrol and maintenance, phase balancing, mounting signs indicating fuse link size, painting and stenciling transformers, etc. Servicing OCRs, sectionalizers and oil switches may be handled this way, or you may report it under an existing unit. We found that it is worth one LA/R when the unit is removed and one LA/I when it is reinstalled.

Of course, travel time to and from the job should be included.

Hours spent attending meetings *should* be included because safety training, employee information, etc., is just part of the cost of nailing up a crossarm or replacing an overloaded transformer. So is "rainy day" time. Whether profitably used on equipment, vehicle and tool maintenance, or on training or used shooting craps or playing poker, it should be included in time charged against putting units in the air (or in the ground).

Verification. It is essential that everyone involved know that each unit reported is easily verifiable and that it *is* verified. Note that TONs and MCOs are reported monthly on a transaction report based on standard utility accounting procedures. CLTs can be checked with dollars collected for reconnects. SUB and CKR can be checked against monthly reports which these inspections generate. CMPs must be recorded in the office under regulatory agency rules. Meter logs to check meters. You may have a report in your company for LMPs, or you may just check this against the number of lamps requisitioned. ROs can simply be counted.

Construction units should be spot checked from time to time against work prints and against work order closing reports.

You don't hire thieves and liars to work for your company so you can dismiss the idea that I'm suggesting your people will cheat on the reporting. It just does not happen. What you are doing by closely supervising the accuracy of E/R reporting is letting your people know that none of them is going to be made to look bad as a result of someone's misunderstanding of the proper way to report or some honest mathematical error. You can't afford to hear the comment, "Well, if we reported E/R units the way they do over in Podunk District, we'd look good, too!" This would kill your program. The workers must be convinced that reporting is fair and honest throughout.

It is not our primary purpose to measure the efficiency of individuals or work groups, as such. It is our primary purpose to measure the effectiveness of their supervision as evidenced by their workers' efficiency. Of course, we don't apologize for measuring individual and group efficiency because our rating system will identify problem individuals and groups, and we must have that information to properly supervise. However, the vast majority of our workers will perform pretty much the same, given identical work conditions, tools, equipment and supervision. Individual differences due to metabolism, orneriness, intelligence, etc. will be reflected, but the overriding controllable factor is *supervision*. Work conditions, tools and equipment are pretty well set by company policies and are generally comparable from one man, group, or location to another. Supervision is not. The possible variations in quality of supervision are numerous. This manager, or superintendent, or foreman is great. Another is lousy. Here we have a lousy manager, a tolerable superintendent, one great foreman and two old foremen looking only for early retirement. One rose in a patch of weeds. How effective can he be?

It has been my experience over 37 years that the measure of a worker's or a supervisor's performance by his superiors is determined largely by his responsiveness on an emotional basis to his superiors. Is he "cooperative"? "Is he a troublemaker"? If the individual tends to inflate the ego of his superior, he is tagged a "good man." If the superior feels threatened in any way by the individual, due to an inherent questioning attitude or due to superior talent made all too obvious, the individual is tagged with various oblique and slighting references calculated to cut him down to size.

Without a scientific means of measuring efficiency, great injustices are inevitable, given the present state of grace of the human race. Evaluation is largely a matter of emotion and general prejudice.

The units previously described give us the scientific means of producing just such a measurement. They are simple, easy to understand. They are few in number. Note that there are actually only nineteen construction units with subdivisions of services, transformers, and conductors. It is not complicated by separate accounting for underground units as opposed to overhead units.

Now that we have identified the units which we desire our force to produce in quantity and quality, let's move on to a discussion of how to "load" the units to produce consistently accurate results.

Develop a measurement for each unit. It is important to understand at the outset that we needn't "load" the units at all to get results which will tell us which individuals and groups are performing well (have good supervision) and which are not. We could simply assign one point to each unit and record the total for each individual or group on a month or on a twelve-month-to-date basis. This would be sufficient to measure relative accomplishment.

But to the extent we can assign accurate average times to each unit we will be able to measure not only relative accomplishment between individuals and groups, but also between actual work accomplishment and an ideal 100 percent accomplishment. This 100 percent should not be based on some mythical super-serviceman or crew of men. Rather, it should be a measure of work accomplishment compared to the best *existing* serviceman or crew. Such servicemen, meter readers and crews now exist in your organization, and because of their consistent outstanding performance under quality supervision their performance is outstanding. Despite all that poor management can do, there will always be some workers who excel despite the obstacles we place in their path. Identify these individuals and work groups and concentrate on them.

Start with a pushover. Meter reading. You know how many dials are read each month in the entire area under your scrutiny. You know how many man-hours are spent on this activity. Divide total man-hours by total meters and you get your first rough measure of the meter reading unit.

Now let's isolate on three or four of the known best readers and evaluate their per unit performance. You will find that it beats your overall average, but not by as much as you expected.

I am aware that some meters may be read by managers, servicemen, or construction personnel. No matter. You would not expect them to read as efficiently as your full-time readers, and that is reflected in the lackluster overall average. But that overall average gives you a limit. You *know* meters can and should be read faster than that. With the evaluation of the sample of "best" readers, you begin to get closer to an accurate figure which should be applicable to *anyone* reading meters. We know of no valid excuse why non-full-time meter readers should read slower than full-time meter readers.

There is still one fly in the soup. We know that a reader cannot be expected to read a "riding" (usually rural) route and produce as many dial readings in a day as a man reading a route consisting of "walkers" (usually close-spaced urban meters).

Again, we must isolate. Test a 100 percent, or nearly 100 percent, riding route and a similar walking route with "best" readers. I have found an almost exact 2 to1 ratio. Whatever your results, reduce the per-meter figures to reasonably round numbers. This is important at this stage because you will be doing all calculations by hand. Long, trailing digits after a decimal are an invitation to disaster in the realm of accuracy. Also, they tend to cloud your easy grasp of the overall picture. A simple .02 hr/ meter is easily recognizable as 50 meters/hour (.01983726 doesn't mean a thing). After the system is established and computerized you can get as snotty as you like with long decimals, if that sort of thing holds great fascination for you. But I would not advise it even then, because it tends to indicate that you're a lot smarter than you have any right to be. It is more important to produce short simple loading values that *work* when applied to live human beings than it is to claim that anyone can know to eight decimal places how long it takes to read a meter. No one does. No one ever will. We want to produce a simple little measure of performance for some of our fellowmen which will give us a reasonably accurate indication of how they are doing in their jobs, not a mathematical monster to bow down to and worship.

You now should have one simple number to apply to walking meters and one simple number to apply to riding meters. Test several routes, some predominantly walkers, some predominantly riders, and some mixed. Apply your unit loadings and compare with actual reading times recorded.

It is important to point out at this juncture that once readers learn you are checking their results, they don't take long to figure out that a "rider" credits more than a "walker," and "walkers" begin to disappear and "riders" proliferate. Therefore, early in the program while you're still smarter than they are, establish an accurate count of walkers and riders on each route by sending an unbiased observer along to properly record them. Even then, you will find that many of the meters honestly recorded as riders should be walkers because a meter reader would give his front seat in Hell before he would walk to any meter that he can find any excuse to ride to. You have already uncovered a built-in inefficiency which can easily be corrected by proper management.

Since maintaining an accurate division between walkers and riders presents a constant headache, why not just count meters and avoid all the confusion? Because if we did that we would inflict a grave injustice on those areas that actually do have an inordinate percentage of riders. To avoid this is of sufficient importance to take on the nuisance of continually monitoring the ratio.

Once a year is often enough to check the walker/rider ratio. Do not make it a monthly torment.

After establishing the walker/rider ratio and assigning a

loading to each and then applying these test loadings to sample routes of various ratios, you will see that you have ample information to fine tune your loadings to final figures which you will find to be valid for years to come.

The results will be tallied every month, and after twelve months you can begin a tally of twelve-month-to-date figures which will erase effects of varying weather conditions throughout the year. A chart of twelve-months-to-date figures will provide a clear indication of how each individual or group is progressing. From such figures, I have spotted men who had become seriously ill without telling anyone. Also, men who were having serious family problems. Of course, it is gratifying to see upturns resulting from actions taken by management to improve performance. Good supervision replacing mediocre supervision will be clearly indicated. But most helpful is the indication of gradual deterioration or improvement in individuals or groups which calls for investigation to pinpoint the cause.

This value of comparison over extended periods of time should not be screwed up by constant tinkering with the loading of units of work. Get them right initially; that is, very early in the program. By the time I had the second month's report in hand for my entire division, I was able to complete my fine tuning and have not found it necessary to tinker with the loading since. That statement applies to all 33 loadings, not just meter reading.

Which brings us to the establishment of loadings for the eight serviceman units.

This is done by following the same basic logic in establishing meter reading unit loadings. Isolate the unit and test it, using one of your best workers. It may be practical to have him spend a week, or at least several scattered days working on only one unit. Other work units may be tested by having a good worker tabulate the time required every time he performs a task. In this case, it is necessary to use results from several workers. Some work units allow for a limit test by taking a month's output and comparing it with the known percentage of the involved workers' time spent on that activity.

Indispensable to accurate assignment of loadings is the advice of several experienced supervisors who have actually done the work involved and are not about to be snowed by anyone.

First, let's get an estimate from such wily supervisors of what percentage of the servicemen's time is spent during the year on each of the eight serviceman units and the two meter-reading units. What percentage on preventive maintenance and what percentage on construction type units. Now we have available records to determine how many of each of the work units were performed during the previous 12 months. We know how many turn-ons, meter change-outs, collections, substation inspections, complaints, meters read by servicemen, relampings, routine orders prepared and recloser/capacitor installations inspected. We know how many hours were reported on preventive maintenance and can pretty closely estimate how many hours were consumed on construction units. Now we have this thing where the hair is short. There is only 100 percent. The temptation to claim that each man spends 100 percent of his time on each unit just won't wash. The time spent on each unit, together with the time spent on the other categories must be fitted into just *one* 100 percent total.

The percentage of time assigned to each category is calculated in hours. Into this we divide the total unit output in that category to produce a man-hour per unit loading for test purposes.

This loading is compared with the loadings we got by running tests on individual "best" servicemen. Add a heaping helping of common sense and experience and our tentative final loadings emerge.

Now, we apply these loadings to a small experimental group of servicemen to see if they stand the test. We should not produce rating over 100 percent for anyone, of course. If no one scores higher than 25 percent, we know our loadings are too light. The ratings scored should fit the observed performance of the individuals pretty closely. The integrated rating of the entire group should be within the limits of reason. If any of these parameters is violated the loadings should be fine-tuned and reapplied until they make sense.

We can now apply them to the entire division for a couple of test months and complete the fine-tuning process.

And so we come to the most difficult task, assigning loadings to the 23 construction units.

Let's first cast out a couple of methods which will probably be advanced. Man-hour values assigned by your company, probably in 1926, to a myriad of units, including some doozies that haven't been in use for fifty years. There are too many of them to be of any use in our simplified program and their accuracy is very doubtful. They are of no value to us.

Next, why not divide contractor unit prices by a man-hour dollar value and produce instant man-hour loadings? On the surface, this looks good, but before you heave that sigh, compare two or three contractors' unit prices and you'll find that one is heavy on conductors, another on poles, another on pole-mounted equipment, etc. You then realize that they are still using old, old, man-hour figures to produce unit prices. That's why they scream about losing money when you hand them a job to do which happens to involve a majority of units from the lightly loaded section of their price list. And they just smile and keep silent when it goes the other way. In the long run they depend on getting a mix of units which will insure a reasonable profit overall, even though they consistently lose on some units and make a killing on others.

This does illustrate, however, that a poorly loaded set of units can produce adequate results over a long enough time span if the totals consistently match reality to an acceptable degree.

We can afford to abide a little "slop" in *our* unit loadings but not as much as you will find in most contractors' unit price schedules.

So, do we assign some poor devil with a stopwatch to observe construction in action and record "actual" times spent on each unit? I would sooner try to stab an oyster. Or make a drawing of fifty puppies at play. No way. The subject is too complex. Too many variables. Too much overlapping activity.

The proper way to do it is to follow the same logic we used in establishing unit loadings for meter readers and servicemen. Isolate. Field test and time significant quantities of work. Establish tentative loadings. Apply these loadings and fine tune them.

I asked one construction unit to note on each routine order or small work order how long it took to complete from the time they left the service center to go to the job (or from the time they left a previous job to go to the subject job) until they packed up and left the finished subject job. They did this for about six months.

With this information in hand I had the following facts to work with. I had the actual man-hours spent on each job. I could easily tally the simple work units for each job. And I knew that these people were representative of about the best we had at the time.

Naturally, I began by sorting through the job drawings to find those involving only one unit. The sum of all the man-hours on those jobs divided by the sum of all the units gave me a tentative loading for each of the very few units I could thus isolate. And I knew that this figure had to represent a top limit, because a single unit job is unavoidably a low efficiency job. But it was a start.

Next, I sorted out the jobs involving only two or three units, which included my solitary first tentatively loaded unit. You will immediately guess that I found some three-unit jobs which took little longer than my one-unit job. Not to panic. That's to be expected. Let's say our one-unit job was to install a lightning arrester and it took 1½ man-hours. A two-unit job involving installing a lightning arrester and a cutout took 2 hours. Common sense tells us that the proper times to assign to a lightning arrester and a cutout should be very similar. So a tentative loading of one man-hour is indicated for both units. The upper limit of 1½ man-hours recorded for the one-unit job bears out the reasonableness of this assumption.

In this way we continue to peel off all of the more easily isolated and identifiable units and assign tentative loadings to each. For instance, it's relatively easy to isolate and evaluate installation of an ancohor, a guy, a service, a COL or a street light.

With this many units tentatively loaded we are ready to analyze some bigger game—some multi-unit jobs involving "known" units plus one "unknown" unit. The total value in man-hours of the known units subtracted from the total man-hours charged to the job leaves the remainder as the tentative loading for the unknown unit. If this produces an unrealistic figure, it's time to reexamine our tentative loadings for the known units.

Bear in mind, when I say "*a*" job, I mean the average values derived from a goodly sample of like jobs. So when you apply the last test I described, you will find that you don't produce wildly unrealistic figures for the unknown unit.

When you analyze a sampling of larger transformer installations or large service installations you *will* get significantly larger man-hour figures than you got on small transformers/services. Because it does take longer. And your figures recorded from actual experience will bear that out.

This logical process is followed as far as practical. Stop just before you begin to feel silly. Stop before you begin to push it past the point warranted by the facts at hand. As soon as you find yourself guessing, quit.

You're still a long way from a complete, accurate listing of unit loadings for all 23 units. But you do have in hand a sizable amount of solid scientific data to help you corner the rest of the units. You are well on the way to a final, satisfactory conclusion.

Now, let's use a different tack to peel off some more units.

Take the guy-wire protector. Why include such an insignificant item at all? Well, on our system it is a unit of property, reportable on work order closing reports. It is also installed in small quantity by servicemen as an isolated task in locations where trouble has been reported as a result of cattle rubbing a guy wire. Now, loading this troublesome little unit is not possible by our methods described to this point. It's not going to break our backs whatever loading we assign, within reason, because the total man-hours spent on this unit is bound to be highly insignificant compared to the whole.

This is one unit that lends itself to evaluation by the stopwatch method. How many can be installed in an hour? Quite a doggone few, you'll agree, if all the work is confined to one site. But we know that many are installed by servicemen (as previously mentioned) in a setting involving a special trip to install only one protector. It is sadly true that there have probably been instances where a serviceman drove 20 miles to a site, installed one protector and then drove 20 miles back. Then in the afternoon, he drove back to the site to perform a turn-on next door to the shiny new protector. But we ain't going to load units based on this kind of stupidity.

On the basis of observation and experience we settled on a loading corresponding to several protectors an hour.

Comparison of units which consume similar times. No matter how much time it takes to install item A, we know from experience that item B takes an equal amount of time. Well, maybe not precisely equal, but fascinatingly close. And item C, not a nickle's worth of difference. And item D and item E. All heavy-use items. All similar in purpose and likely to be used in quantity on any one job. If we know how long it takes to install 1000 of these items in whatever mix, we can establish a unit loading for each item which if applied to the next 1000 of these items we install in whatever mix will produce a total calculated time very close, indeed, to our original measured total time. So we feel justified in lumping all these items under the "lead" item and reporting them accordingly. For instance, we described a whole family of items under the lead item, Pole Top Pin, or PTP. A rack (any size) is not reported as a rack, it is reported as one PTP.

We install one two-pole steel platform every other year. We say it takes four times as long as a cluster mount. That jibes with our experience. But what if we're wrong? So what? It's a reportable unit of property so we must include it, but it does not deserve much of our time or trouble to produce an unassailably accurate loading for such a seldom-used item.

Ground. One complete installation. Now here, we cheated. I instituted a contractor grounding program to ground all equipment–carrying poles some years ago. We've installed tens of thousands of grounds. We have the records. We know how many man-hours it took. We feel that company personnel should match this performance. Simple division.

Now it's time to stop and look around. What have we accomplished so far? Where do we yet have to go?

Well, we've just about cornered and raped every little isolated, defenseless unit and we've crowded a lot of other little units into groups composed of units which we have more or less arbitrarily defined as equal in time value. We still have no handle on what that time value is. And we have a plan to determine loadings on "unknown" units used on jobs with "known," that is, tentatively loaded units. Many transformer installation jobs can be found involving only the transformer as the "unknown" unit. The same is true for oil-circuit recloser, capacitor and disconnect installations.

We have now produced tentative loadings for 13 of our 23 units. We have applied these tentative loadings to numerous jobs and fine-tuned them to produce "initial" loadings for these 13 units. The initial loadings will be used in our full-scale, two-month dress rehearsal.

Let's list all 23 units and note by asterisk the units we have provided with initial loadings.

POL	PTI	*COL	*DISC
ARM	DE	*GWP	2/0
PTP	*GRD	*50	4/0
*ANC	*LA	*75	750

*GUY	*2/0S	*OCR	CBL
CLS	*4/0S	*CAP	

What's left? You can see that I have avoided poles, all conductors and conductor support items, and cluster mounts. And you notice how quiet I've been on the subject of removal and transfer unit loadings. I'll get there, eventually.

Let's knock off that **cluster mount**. Relatively few installed. The loading can be determined by closing in on it from two sides. One, simply observe and record how long it takes to install a few of them and take the average. Two, treat it as an "unknown" in combination with previously identified units. We have the loading for little and big transformers which we determined by treating them as unknowns or single-phase installations. We will want to use this loading for each of the three transformers installed in a three-phase bank on a cluster mount. Now you know that there is probably a perceptible difference in time actually required to hang one transformer on a single-phase installation and to hang one transformer on a three-phase installation. But we choose to ignore any such difference, whether more or less. We will doggedly apply the loading from the single-phase installation to each transformer of the three-phase installation, set the cluster mount up as an "unknown" and assign any additional leftover time to it as its loading. The loading we derive should be in the ballpark with the loading indicated by the observed timing method.

Poles. Let's talk about poles. There are two things which can foul up construction scheduling. Right-of-way and rock. The right-of-way problem can be solved by building an expert team of negotiators and give them plenty of lead time. The rock problem can be solved by separating pole and anchor installation from the rest of the construction program, give it to specialists and give them plenty of lead time. Except in rare emergencies, I don't let any company employees spend their time and energy banging on rock holes. It used to make me sick watching the old-time crews stand around while one or two men beat on a rock hole. So I got with a contractor and had him put together

a three-man pole and anchor installation crew consisting of one knowledgeable old broken-down lineman and two pick-up local appleknockers. They work for a week in one district, then move to another. We have kept a minimum of two crews busy since 1958 and have had as many as five crews going at once. We have five districts. They have experimented with all types of rigs, but have the best luck with a flatbed truck, hydraulic A-frame, and truck-mounted compressor. The program automatically forces our superintendents to plan their work. The crew's work is planned so that they work in a limited area each day, not having to travel far between jobs. Our company construction unit's work is planned the same way, so there is no lost motion as a result of "You mean you have two work units, one contractor and one company, visit each work site, no matter how small the job?" Yes, we do. And it's the most economical plan. When we hit a bad rock hole, or a water hole, it slows down only the cheap contract pole-stters. Nothing slows down our expensive company forces.

If you are in an area where rock is a problem, you know what I'm talking about. If you don't have any rock then you don't have to contend with the rock delay problems and you have my congratulations. But I still believe you will find that a cheap pole-and-anchor contract crew will pay off because of its obvious economy. Highly trained company employees shouldn't be wasted on low-skill work, and most important, the automatic forcing of efficient job scheduling.

Under our plan you can see that loading the pole unit involves far less time per unit than it would if we dug our own holes. The only time our people set a pole—and that's very rarely—it is set in a hole previously dug by a contractor. Determining a time loading for our pole unit is a relatively simple matter, since we have eliminated the highly variable element of rock. The same thing applies to the way we set anchors. The only time our company folk set an anchor, the hole is already dug. That's how I was able earlier to include anchors in the group that would be easily isolated and evaluated. Our poles can be evaluated in the same way.

You lucky son-of-a-gun, you say you got no rock, but you install poles and anchors with company forces. O.K. Unit loading is no problem for you. Just hit it from two angles, the isolated one-unit job and the job where the pole (or the anchor) is the only "unknown" unit. Woe to the unfortunate who has rock and is condemned to dig it out with company forces! You have two work units, pole and anchor, which include the basic time to install the unit where rock is not involved plus the unknown, widely fluctuating, nonverifiable time required to fight the rock. Any inefficiency can be blamed on rock holes. You have injected an item over which you have no effective control. That's all it takes to lose control altogether.

Naturally, I would advise you to accept my plan and turn to a contractor. But I know this may not be possible, or feasible. Also, you may just disagree with me. You know your situation better than I. You may be right. Maybe I'm full of prunes. But you will agree that it injects a disturbingly undefinable, uncontrollable element into a business where it is of primary importance to eliminate just such factors, or to reduce them to an irreducible minimum.

So what should you do? If I had your problem I would proceed to load the pole and anchor units on the basis of dirt holes and require the construction forces to report the actual man-hours consumed on each rock hole requiring the use of an air compressor. In other words, reduce your problem to just the wicked holes by lumping the minor rock holes into the overall average. Set up a spot checking program on the time reports from the wicked holes, to insure reasonable accuracy. This complicates an otherwise very simple efficiency rating system and I don't like it, but understand you've got me dealing with people who like to dig in rock. I'm limited in the degree to which I can be helpful to you, dummy.

So much for poles and anchors. Bet you thought that would be hard!

The remaining two nonconductor items are **crossarms** and **pole top pins**, with their trailing items. We left them until last because they are admittedly the hardest units to isolate and evaluate without a multitude of other units being installed along with them, making it well-nigh impossible to get a fix on a single item to properly evaluate it.

But now we have initial loading on most of those other items. Only conductor items are unloaded. So we sort through our timed routine orders and small work orders and pick out jobs involving arms and PTP items which do not involve conductor items. Then we pick out jobs which do include conductor items on which we can make a very close estimate of the percentage of the job time which was, or should be, consumed by the conductor items. Subtracting the conductor time and the calculated time spent on "known" units, we are left with the remaining time, which can be divided by the PTP or crossarm items. At this point it may become evident that one or more "known" items is improperly loaded, and further adjustment is in order. But that's O.K. We know that we are moving asymptotically toward an accurate set of loading figures, and we have actual timed jobs to keep us within the limit of practicality.

This leaves only the conductor items. Let's knock off PTI (pin-type insulators) first. We know that the time assigned to this unit should be comparatively small. If we err, let's err toward too little loading rather than too heavy loading. No job is going to be made up of nothing but installing PTIs. They are always an accompanying item. If they weren't a unit of property we could ignore them as a separate item and load them into PTPs, ARMs and/or conductor items. But we want to be able to verify every unit of property installed against those claimed under our E/R (efficiency rating) system. We settled on 10 PTIs per hour. We know that can't be far wrong. And it has worked out well, not producing any distortions in our system, over the six years we have used it.

We're getting close, now. All that's left is conductor and dead-ends. Take **dead-ends.** How long does it take to install one? First, let's recognize that the vast majority of these units involve small and medium size conductors. So what if it takes 20 times as long to install a dead-end on 1,000,000 CM than it does on #2ACSR? We'll just load the big, big wire enough to include anything we cheat a crew out of on big wire jobs by crediting those big dead-ends at the smaller wire loadings. By thus limiting our problem we can concentrate on producing a loading which fits the preponderance of the dead-ends actually installed.

This unit can best be evaluated by observing several jobs on which we find a single lineman busily engaged only in making up primary dead-ends on a pole full of dead-ends and divide his time by the number of dead-ends. Then do the same thing on a pole where several secondary dead-ends are involved. With this actual field data in hand we can apply a little common sense based on long experience and settle on a time loading which may not be as pure as the driven snow scientifically, but it will be very close. Close enough so that this loading will not throw the rest of our system out of whack. Remember, we're going to make it all fit into just one 100 percent based on a generous sampling of jobs. Finally, we'll make it all fit into just *one* 100 percent based on a two-months run of all the jobs done in the entire area.

Now, **conductors.** Even at this late date we look for strays which we can easily cut out of the herd. Obviously, cable is the most vulnerable work unit to attack. We now have enough "known" units so that we can sort out of our field-timed jobs those in which cable is the major factor. The one "unknown." We add up the total time consumed on a goodly sample of representative cable jobs and subtract the calculated time assigned to "known" items, based on our latest initial values and divide that result by the number of 100 foot long cable units involved. The resulting CBL unit loading is then applied to numerous samples to see how well it holds up job-to-job. We fine tune the loading and are ready to apply it as an initial value on our two-month system-wide test run.

Finally, we divide conductors into three categories—small, medium, and large. We have included the labor of tangent, angle and dead-end attachments in previously loaded units PTP and

DE. So all we have left is the actual conductor stringing and sagging, or pulling into conduit. We will not differentiate between "hot" and "cold." It is rare to find a job on which you can string conductor "cold." The same care is usually required for just about every job, or should be. We are seeking the average time required to install 100 feet of conductor, and we want it broken down into three general sizes involving three different levels of equipment and manpower. The average is going to approximate the time loading for "hot" installation. And we won't differentiate between primary and secondary. The conductor doesn't know what voltage it's going to be required to carry once it's in place. It requires equal care to string a secondary conductor through a hazardous route of other conductors, some energized, some bare, over roads, through traffic, etc., etc., as it does to string a primary conductor through a similar maze.

You will perceive that our small-conductor category covers almost all the conductor work most crews will be involved with in the course of a year. It just about covers all their wire work except feeder circuits and heavy secondaries (which is about all there is in our medium-conductor category). The heavy-conductor category includes only those very heavy feeder circuit installations involving special stringing methods.

To establish accurate time loadings for the large-conductor category and for the medium-conductor category, we must record time required on several actual jobs in progress as performed by work units which we consider to be our best. Then apply these loadings to jobs done in the past on which you have reliable information. How long it took to perform *only* the conductor installation procedure. Apply these loadings to a few simple feeder circuit jobs, jobs involving a minimum of extraneous units, using initial loading values for the extraneous units to isolate the conductor installation time. By this method we can determine reliable unit loadings which can be used as initial loadings in our two-month final run.

We have now loaded all our units but one, small conductor, the most troublesome unit to get a grip on because it includes so many variables. Small wire feeder circuits, both short and long. Single-phase and multiphase. Hot and cold. Secondaries.

But we have this evasive little devil in a corner. We have initial loadings for all the other units, so we can isolate this unit for various types of jobs out of our selection of test-run jobs for which we have total time required. And we can compare this unit with the "known" medium-conductor and large-conductor units to see in what range we can expect to find this unit. Of course, we can observe actual times required for conductor installation on several types of jobs, including both overhead and underground. Keep in mind that the loading of this unit should be influenced much more by the run-of-the-mill type job which comprises the vast majority, and less by the exotic, or rare type of installation.

We can also cheat a little on this one and take a peek at contractor prices. The highest contractor conductor price for this range of conductor size will give us an upper-limit loading and the lowest contractor conductor price for this range of conductor will give us a lower-limit loading.

With all of this information in hand, a practical unit loading will emerge which you can use on a sampling of jobs (along with your previously derived initial loadings for all the other units) to test the validity of your choice. Fine tune it, and you are ready to begin the full-fledged, two-month dress rehearsal covering all the work in your entire system. As soon as you establish loadings for removal and transfer.

How does the overall time spent on **removals** and on **transfers** during any one month or any one year compare to the total actual time available for construction work? A mighty small proportion.

Assume for a minute that we are so doggone tired of sifting out these loadings at this point that we just throw up our hands and say "Let's use the install loading for removal and transfer, too!" I wouldn't blame you for being tired of my seemingly endless harangue, but you are too conscientious to consider such mental sloppiness.

The point is, that such a course would have minimal effect because it would involve only a small proportion of the total units. Besides, a lot of the remove and transfer units *are* equal in loading to the install units. And the average of those heavier, lighter and equal is not far from the total of the install loadings. So, to whatever extent we can refine our remove and transfer loadings, to that extent we will enhance our overall accuracy.

It is not easy to find a sizeable sampling of jobs involving only removal units. So we must fall back on that source of limitless expertise and experience to which I referred earlier, the long-time superintendent or foreman who has had actual experience doing this work himself, as well as supervising it. His judgement of the relative times required to install, remove or transfer a unit becomes invaluable. Your considered opinion will enter in, too, of course,

Let's look at some individual items. Take poles (remember, I only install them in pre-dug holes) and COLs. We concluded that there wasn't a nickel's worth of difference in the I, R, T loadings for these two units.

The transfer loading for 12 of the items is obviously equal to the sum of the install and the remove loadings. That's because transfer involves removing that item and reinstalling it. These 12 items are: ARM, PTP, GUY, CLS, PTI, DE, LA, 50, 75, OCR, CAP and DSC.

Besides poles and COLs there are five units that we agreed take an equal time to remove as to install. These are: ARM, PTP, CLS, PTI and GWP.

To avoid getting lost, let's tabulate all the units and the relative loadings, based on the unit's install loading as unity, to the point we have already progressed.

	I	R	T		I	R	T
POL	1	1	1	COL	1	1	1
ARM	I	1	2	GWP	1	1	
PTP	1	1	2	50			I + R
ANC				75			I + R
GUY			I + R	OCR			I + R

	I	R	T		I	R	T
CLS	I	1	2	CAP			I + R
PTI	1	1	2	DSC			I + R
DE			I + R	2/0			
GRD				4/0			
LA			I + R	750			
2/0 S				CBL			
4/0 S							

We see that whatever loading is assigned to POL/I, the same loading applies to POL/R and POL/T. Whatever loading is assigned to ARM/I, the same loading is applied to ARM/R and twice that amount to ARM/T. We know the loading for GUY/I, but we haven't yet determined what GUY/R should be. But we do know that GUY/T will be the sum of GUY/I and GUY/R.

Let's whittle away some more of our ignorance. Eventually we will get that chart all filled in.

Four cheap shots. In defining some of our units, we pointed out that no credit was due for removing and/or transferring that unit. No credit for ANC/T. No credit for GRD/R or GRD/T. And no credit for GWP/T. That knocks off four more of our 23 X 3 = 69 value slots.

We concluded that on nine items it was reasonable to assign one-half as much time to removal as we had to installation. Those items were LA, 2/0S, 50, OCR, CAP, DSC, 2/0, 4/0, and 750.

Now our chart looks like this:

	I	R	T		I	R	T
POL	1	1	1	COL	1	1	1
ARM	1	1	2	GWP	1	1	—
PTP	1	1	2	50	1	.5	1.5
ANC	1		—	75	1		
GUY	1			OCR	1	.5	1.5
CLS	1	1	2	CAP	1	.5	1.5
PTI	1	1	2	DSC	1	.5	1.5
DE	1			2/0	1	.5	1.5
GRD	1	—	—	4/0	1	.5	1.5
LA	1	.5	1.5	750	1	.5	1.5
2/0 S	1	.5		CBL	1		
4/0 S	1						

You can see that we're pretty close to licking this thing. Removal of a guy takes one-fifth as long as installing a guy. Or so it seemed to us. And we concluded that removing an anchor, which usually involves only screwing out the rod, was worth only half as much time credit as removing a guy.

We decided that ten DEs could be removed in the time it takes to install four. DE/R, then, equals .4 DE/I.

We determined that it takes twice as long to remove a big service as to remove a small one, and equal time to transfer a service as to remove one.

We determined that it takes longer to remove a big transformer than to remove a small one, but less time to remove a big transformer than to install a small one. Since we had set a 50/I at 2 times 50/R, it was evident that 75/R should fall in between these two values, or about equal to 1.5 times a 50/R.

Finally, we have reached the last item, CBL.

Cable secondary is usually removed along with other work for which you have already set up, while transferring cable is usually done as a separate job, requiring its own travel time and set up. We concluded that it takes twice as long to transfer as to remove and two-thirds as long to remove cable as to remove 2/0 conductor. We thought installing cable was a little longer job than installing 2/0 conductor and settled on a value of seven-sixths of 2/0. Now you see why we left this unit until last. It is a comparatively infrequent job, so we were reduced to evaluating it on the basis of every observation and comparison we could muster. The loading figures we evolved proved to be fairly accurate. As long as such a seldom-used item is loaded within reasonable limits, it will not mortally damage the overall acuracy of our rating system.

	I	R	T		I	R	T
POL	1	1	1	COL	1	1	1
ARM	1	1	2	GWP	1	1	—
PTP	1	1	2	50	1	.5	1.5
ANC	1	.5 x GUY/R	—	75	1	1.5 x 50/R	1 + 1.5 x 50/R
GUY	1	.2	1.2	OCR	1	.5	1.5
CLS	1	1	2	CAP	1	.5	1.5
PTI	1	1	2	DSC	1	.5	1.5
DE	1	.4	1.7	2/0	1	.5	1.5
GRD	1	—	—	4/0	1	.5	1.5
LA	1	.5	1.5	750	1	.5	1.5
2/0 S	1	.5	.5	CBL	1	2/3 x 2/0/R	4/3 x 2/0/R
4/0 S	1	2 x 2/0 S/R	2 x 2/0 S/R				

By replacing unity in the install column with the previously determined initial loadings we can replace the remove figures and the transfer figures with their corresponding values.

Whatever values you produce for construction, service work, and for meter reading, I recommend that they be kept strictly confidential. I will expound on this at some length in the next section, but it is so important to the continuing validity of the loading values you finally settle on, that I want to emphasize it here and now.

Let me review what we have accomplished so far.

I have defined efficiency and the need for measuring it in the operation of a utility.

I have offered a method of measuring efficiency which I believe is superior to any other method in use today. It is simple and accurate. It provides the information required to evaluate the performance of the individuals and groups of workers, and their supervisors, whose work comprises the bulk of our activity. Each worker or work group need record only a very limited number of units at the end of each day. Five minutes maximum recording time is involved. The accuracy of the reporting easily verifiable against existing company reports and records, promoting confidence in all those involved in the accuracy and fairness of the system. The information derived makes it possible to make pertinent comparisons between performance of individuals and groups and it provides a measure of the performance of others whose work is closely related to those actually being monitored, such as office personnel and engineers.

The method involves determining the hours available for productive work. Then we must evaluate the units of work performed by establishing time loadings for each of the 33 defined tasks. The units, converted to hours, divided by the available hours gives us a ratio of production versus time available for production. This is our efficiency rating, or E/R.

I defined available time in some detail.

I then defined each of the 33 task units.

Then I described my method of loading each unit. This method involves the following methods of attack.

1.   Compile a large sampling of test jobs, routine orders and work orders done by "best" workers.

2.   Time experimental jobs involving only one item, or a minimum of extraneous items.

3.   Utilize the vast experience of old-line supervisors who have actually done all classes of work.

4.   Isolate an "unknown" unit in a sampling of jobs predominating with that unit but which also include other units for which you have "known" initial loadings.

5.   As in the case of grounds where we had records on contractor installation of thousands of units, we easily determined an accurate loading.

6.   Use contractor unit prices to establish high and low limits on conductor, and to compare times involved in transfer and removal compared to installation.

7.   I discussed loading rock pole and anchor holes and my prejudices that poles and anchors should be installed by contractor forces, not by company personnel.

8.   In the case of servicemen, establish what percentage of their available time is spent on each task. Compare the resulting hours on each category with a month's output of units

in each category. Compare the loadings you derive this way with loadings derived by other means.

9.   Use system-wide overall known times corresponding to total units, as in the case of meter reading.

10.   After initial values are determined, apply them to a sampling of representative jobs and fine tune them as necessary.

11.   Compare initial loadings between units for practicality. Compare initial loadings of remove and transfer with install for a particular unit and with other units.

12.   Run a two-month, system-wide trial using your initial loadings and perform final fine tuning as indicated.

We have our gun loaded, cocked and aimed; now we're ready to pull the trigger!

INSTITUTE THE PROGRAM

You are a meter reader, or a serviceman, or a second-class lineman, and you hear that "they" are going to start "checking on you." What are your feelings?

"If they do that, they'll get a union, for sure!"

"If they try any of that crap, we'll ask the union to strike!"

"If they think they can get any more work out of me, just let 'em try. I'm running with my tongue out now."

"They're going to rate us. *They* never did a day's work in their lives. What do they know about measuring my work?"

"You know what's going to happen. They'll see to it that all their pets show up good, and everybody they want to get rid of will show up bad."

"Now we'll never get any work done for making out all the goddam forms."

These genuine concerns must be effectively dealt with if we are to have any hope of successfully instituting this new program. How can that be accomplished?

Well, it's already been accomplished to a limited degree if you have judiciously chosen the individuals and work groups to provide experimental data. When they are asked to record times for various jobs a great deal of care should be exercised in taking the time to fully explain what is being done and sell them on the idea. Show them some such plan is *required* by some unions, by some regulatory agencies, and by some investing firms. Better to institute our own simple, effective plan before some "outsiders" come in and mandate some whore's dream that will keep us all bound up in our accounting underwear for the rest of our lives. Prove to them that it is in *their* interest to have such a program. Remember, you're talking to your "best" people and they know that they are the best. Don't they want that fact scientifically verified to improve their relative position with the rest of the force who can only hope to show up as average?

Emphasize the fact that the primary purpose is to measure the effectiveness of their supervisors—not them. Explain that ratings of individuals will never be divulged except to their supervisors, and that ratings of groups will be published monthly for everyone to see. Tell them that the norms will be based not on some artificial ideal to try to get more out of the workers but on the existing actual performance of groups like theirs. And that's why you need their enthusiastic cooperation in producing experimental data which they can trust, because they, themselves, produced it.

You will be surprised at how readily these workers will join in. I'm assuming you have a long record of believability behind you. They've heard you say, "Hell, no, we won't do it," and then explain why. And on other occasions you have thanked them for suggestions and they've seen those suggestions implemented.

They know you're a square shooter and you don't have to overcome a background of suspicion and animosity. If you've got those problems, you have started E/R too soon. You need first to establish a program to improve employee relations.

If you are believable, they will believe you, and despite the usual wisecracks, you will have a solid core of support to build on.

Of course, you will have previously done this selling job on the limited number of supervisors involved in the experimental test runs. Now, it's time to do this same selling job on all the supervisors, using your success with the first group as your wedge to pry your way into their hearts. Here is your toughest group. They're already P.O.'d that you started with some other group. Let's offset that by asking their help on further experiments with their own workers.

A large part of the selling job consists of explaining the units to all involved and at the same time emphasizing the simplicity of the system. They'll have lots of suggestions for changes and some of them you may decide to accept. But as soon as you propose something simple the very people who so dread more paperwork want to complicate it. You must insist on keeping your list of units as small as possible. Admit that under this system an individual or group can easily score only 50 percent for any one day, or 150 percent for some other day. That's to be expected. Units are not recorded until they are complete. A crew may spend all day stringing wire and leave till tomorrow clipping in and, as a result, record few units completed for that day. But tomorrow, they'll make a helluva comeback. The system is designed to give a very accurate measure of a year's work and a pretty good measure of a month's work. But don't expect it to reflect an accurate measure of performance on a one-day basis. That's the price we pay for keeping it simple.

Let's go back to confidentiality for a moment. We've discussed this thing with lots of people. We've even asked a limited number of people for advice on loading individual items. So what's confidential? Aren't there lots of people who have a good idea what the key is? Can't that information be used to calculate any reports of units by an individual or a group?

No. No matter that several individuals give you lots of advice on loading units, they know you're getting advice from others, too. And you never let them know what figures you finally settle on. In my company, I'm the only person who knows the key. Anything two people know is no longer confidential. All the employees know that only I know the key and this little bit of mystique has helped maintain their confidence in the whole system. When I die, they'll find the key in my locked desk drawer. As soon as they find the key to my desk, which I have hidden ingeniously.

Now all the soft-soap is over. Time for the nut-cutting. As always, management comes to the point where it has to announce

"This is where friendship ceases and Hell begins. Now *do it!*"

Each location with one or more meter readers and/or one or more servicemen is issued a rubber stamp which is used to print on the back of his time sheet the meter reader/servicemen units.

Each location with a construction force is issued a rubber stamp which is used to print on the back of the crew time sheet the construction units.

They look like this:

POL				TON	
ARM				MCO	
PTP				CLT	
ANC				SUB	
GUY				CMP	
CLS				RMW	
PTI				RMR	
DE				LMP	
GRD				RO	
LA				CKR	
2/0 S					
4/0 S					
COL					
GWP					
Δ50					
Δ75					
OCR					
CAP					
DSC					
2/0					
4/0					
750					
CBL					

At the end of each day the completed units are recorded. This requires about two minutes for a serviceman or meter reader and five to ten minutes for a construction unit. That is, for each one- or two- or five-man (whatever) size unit that spent the day working together. At the end of each month a clerk compiles the total units for each serviceman, each meter reader, and for the whole construction unit at that location. This is recorded on a form shown as Figure 4.

My division has five districts, so I receive one report from each district each month. The construction units are reported for each district on a blank sheet of paper stamped with the construction unit rubber stamp and the SVM/MR rubber stamp (for the few units of that type done during the month by the construction force) with the totals of the monthly units filled in. Occasionally, servicemen or meter readers do a few construction type units and that is reported by using the construction stamp on the back of the monthly SVM/MR form.

Note that this monthly form and the stamps are pretty crude compared to the usual division or G.O. report forms, or compared to computer print-outs. That's no accident. I don't know about your people, but Kentuckians just don't trust slick forms and they detest computer print-outs. By using homemade looking stamps and forms we tended to de-fuse this inherent aversion to yet another damn form.

In this same vein, I record the preventive maintenance hours in the left margin of the SVM/MR form and the hours credited to construction in the space just before the words "Div. Use" for each man.

I do all the calculations, myself. It takes about a day a month, along with my regular work for the day. I convert the units to hours and the total credited hours for each man on Div. Use line under total. This divided by his available hours gives his E/R rating for the month which is recorded on the Div. Use line in the Avail. Hours column.

The total hours credited to servicemen is divided by the total available hours to get the E/R for that district's servicemen for the month. Same for meter readers. Same for the district construction force.

EFFICIENCY RATINGS
SERVICEMAN AND METER READERS

NOTE: DO NOT SUPPLY TOTALS

MONTH _____

SIGNED _____ DISTRICT MANAGER

% P.M.	NAME	TON	MCO	CLT	SUB	CMP	RMW	RMR	LMP	RO	CKR	TOTAL	AVAIL. HRS.
1	TOM JONES	UNITS											HRS.
	CRD.HRS. CONST. DIV. USE	CRD. → HRS.										→ CRD. HRS.	% E/R
2	DIV. USE												
3	DIV. USE												
4	DIV. USE												
5	DIV. USE												
6	DIV. USE												
7	DIV. USE												

SERVICEMAN

8	DIV. USE										
9	DIV. USE										
10	DIV. USE										
11	DIV. USE										
12	DIV. USE										
	TOTALS										
1	DIV. USE										
2	DIV. USE										
3	DIV. USE										
4	DIV. USE										
5	DIV. USE										
	TOTALS										

METER READERS

Fig. 4

The district totals are tabulated to give the division total. I keep the 12-month totals on another homemade form.

Every month I send each district a hand-written tabulation of the results (Fig. 5).

Now that our people have been into this for five years and have accepted it, I wish we could computerize. That would be so easy, as you can see. All those calculations done automatically. But until the system is accepted company-wide, forget it. There's no room at the computer inn for a crazy division engineer with another of his cock-eyed schemes. Even though this crazy scheme has produced by far the lowest cost per customer of any division in the company and our relative position has shown steady improvement since its inception.

The district total of each SVM unit is compared to the available reports to insure accurate reporting. This quickly shows up instances where units are inadvertently reported in the wrong column, or where a new serviceman is reporting on the basis of his misconceptions of the unit definitions. An occasional call to the district to question reporting lets them know that I mean business about accuracy.

By the way, the percent of each serviceman's time spent on preventive maintenance and the district percentage spent on preventive maintenance is also recorded on the monthly form and reported back to the district.

So that I can maintain a constant control over construction force reporting, I receive the daily time sheets from one of my five construction forces. On the back of the time sheet is the E/R units claimed for that day. It takes about one minute to calculate the E/R percentage for that day. In this way, I can pick up inadvertent inaccuracies in the reporting of units and hours. If one day yields 150 percent, the next had better be 50 percent, or 40 percent. No three days should average more than 100 percent. If they did, I would begin to worry about my time loadings. But they don't, thank goodness.

12 MO THRU OCT 80												
	← CRD	SVM AVL	→ %	← CRD	MRS AVL	→ %	← CRD	CONST AVL	→ %	← CRD	SMC AVL	→ %
DISTRICT 1												
12 →SEP 80												
OCT 79												
OCT 80												
12 →OCT 80												
DISTRICT 2												
12 →SEP 80												
OCT 79												
OCT 80												
12 →OCT 80												
DISTRICT 3												
12 →SEP 80												
OCT 79												
OCT 80												
12 →OCT 80												
DISTRICT 4												
12 →SEP 80												
OCT 79												
OCT 80												
12 →OCT 80												
DISTRICT 5												
12 →SEP 80												
OCT 79												
OCT 80												
12 →OCT 80												
DIVISION												
12 →SEP 80												
OCT 79												
OCT 80												
12 →OCT 80												

Fig. 5

EVALUATE THE DATA

It would be a crime to stop at this point. Just glow in the knowledge that we have successfully foisted onto our work producers yet another form to warm our bureaucratic egos. No. We must keep before us the purposes we enumerated at the outset and see those purpose served.

We have done a helluva lot of work to produce a system which will cause a minimum amount of work for the workers involved and which will produce results consisting of as few figures as possible to make evaluation as easy as possible.

Now we have those results. What do they tell us?

The first thing we notice is how well each individual and group did compared to the ideal 100 percent. For instance, the 12-months-to-date E/R for my division for all the servicemen, meter readers and construction forces combined is .70. Immediately, this suggests a look at how each district is doing compared to the division. The districts scored 68, 73, 65, 71 and 73. Two districts out front and one lagging 8 points behind. Why?

Next, let's see what each classification of workers contributed to the .70 overall rating. We find SVM = .61, MRS = .78 and CONST = .75. Meter readers away out front. Construction close behind. Servicemen in the cellar 17 points down from the meter readers. Why?

We seem to have a problem in the serviceman area. (We do.) While the district construction forces ranged .72, .78, .74, .78, and .74 for a total of .75, and the meter readers ranged .85, .70, .68, .85, and .83 for a total of .78, the servicemen ranged only .56, .69, .49, .61, and .68 for a total of .61. Why can district 5

161

servicemen produce at .68 while district 3 produces at only .49, 19 points behind?

Let's look at last month's results. Individual servicemen ranged from .14 to 1.14. Meter readers from .61 to 1.17. District construction forces from .67 to 1.0. Does this raise any questions in your mind?

The twelve-month figures on individual servicemen and meter readers and on the five district construction forces have remained relatively constant over the years. *With exceptions.* A few individuals have gradually dropped. A few have improved noticeably. All of our construction forces have shown gradual improvement. Our division has shown very slow gradual improvement. Why?

It is only on very rare occasions that an individual or construction group scores over 100 percent. This is to be expected. If our ideal 100 percent is realistic, we should find an above-average worker or construction group on a good month breaking 100.

What about that lone serviceman who scored .14?   You guessed it. He's being promoted to manager and removed from the E/R system. The next lowest is .33 and if he doesn't watch out he'll be promoted, too! Well, this *is* a utility company, you know. But despite this sort of aberration we must plod right along trying to do what we know is right in managing our manpower in the face of sometimes bewildering actions from above.

Let's take a look at our top ten servicemen on a twelve-month basis. We find one .91. One .87. One .86. Two .85's. One .79. One .78. One .76. One .75. One .74. And one .73. This demonstrates that we have a sizeable group capable of outperforming the average on a consistent basis.

What pattern do we get from our 10 top meter readers? One .91. Three .88. Two .87. One .86. Two .85. One .84. One .75. Notice the difference?

During the past year six construction groups have scored over .90 or over for a single month.

One servicemen maintained a twelve-month average of .91. Four others, .85 or better.

One meter reader maintained a twelve-month average of .91. Eight others, .85 or better.

These three facts validate our 100 percent ideal in each category. Just as importantly, they validate the lower scores recorded and tell us that we can pinpoint our trouble areas with confidence on the basis of E/R ratings.

E/R measures the effect of bad weather on each class of work.

E/R measures the effect of economic recessions.

E/R measures the effect of replacing personnel.

E/R measures the effect of reorganization of manpower.

E/R measures the need for additional manpower and gives one a scientific measure for denying additional manpower in the face of emotional appeals.

E/R points with an icy finger at areas which are overstaffed.

E/R points out areas in which changes in work methods are indicated and measures the effect of improvements as implemented.

So we see that our E/R rating system provides us with a reliable measure by which we can make comparisons.

We can compare:

Individual vs. individual

Individuals and groups vs. an ideal 100 percent

Work class vs. work class

Entire district vs. entire district

One work class vs. another

Work class in one district vs. work class in another district

Individuals and groups over a period of time

Effects of organizational changes

E/R provides checks on its own validity throughout its entire range of measurement.

E/R provides checks on the accuracy of the reporting submitted from the field.

Probably the most important result of our E/R system is in providing each individual and group with the assurance that he, or it, is being judged on the basis of a scientific, fair, unbiased scheme that keeps the performance of the individual or group in proper perspective in front of his supervisors, at the same time evaluating the effectiveness of that supervision. The message that goes out with each monthly report is, "Yes, somebody *does* care."

E/R TO MEASURE OFFICE ACTIVITY

I mentioned before that E/R can be used to measure the effectiveness of other employees besides meter readers, servicemen and construction forces. The information gleaned in the E/R reporting process can be utilized to measure the activity level of each local office.

To what purpose? Well, if your measure of need for manpower in each office is based solely on the number of customers served out of that office, then I assure you that you are breaking some folks' backs while others are taking a free, comfortable ride. Customer count is a perfectly awful way to measure office load.

The customers in one office will generate three to four times as much office activity as the customers in another office. Actual office activity should determine manpower needs. For instance, an office located adjacent to an army camp will be overrun with turn-offs and turn-ons, whereas an office located in an agricultural community will have a turn-off only when a barn burns and a turn-on is accompanied by a parade down Main Street for the first new family to move into town since the preacher ran off with the organist and they both had to be replaced from the outside.

Manpower should be carefully sized to fit the work load. Using customer count as a criteria is a lazy, sloppy, inefficient way to sidestep responsibility. It is a lot easier than digging out the facts necessary to make an accurate determination of real need.

How can E/R help?

Office activity is directly related to three of our E/R service-man's units, TON, MCO, and CLT. Four other serviceman units contribute to office activity, or load. They are CKR, SUB, CMP, and RO. Not nearly as much as their time loadings would in-dicate, however. By experimentation I have found that they should be factored into the determination of office activity at one-tenth of their time loadings.

Meter readings, relamping, preventive maintenance and construction units installed by servicemen are elective tasks which should not be considered since they do not generate sig-nificant office activity and they will vary widely in amount from one office to another.

All office activity is generated by the first seven units. When we take the hours spent by servicemen on the first three units during the past twelve months and add one-tenth of the hours spent on the next four units we have a factor which can be com-pared with similar factors from other offices to establish relative activity. This, of course, is known as the theory of relative activity!

Take an example. Office No. 1 spent 867 hours on TON in the past twelve months, 332 hours on MCO, and 70 hours on CLT, for a total of 1269 hours. 155 hours were spent on CKR, 285 on SUB, 160 on CMP, and 382 on RO, for a total of 982 hours. 1269 plus one-tenth of 982, or 98.2 equals 1367 hours.

Office No. 2 had 1566 TON, 425 MCO and 505 CLT, for a total of 2496 hours. 160 CKR, 635 SUB, 90 CMP and 790 RO, for a total of 1675 hours. 2496 plus one-tenth of 1675, or 167.5 equals 2664 hours.

Office No. 1 has 7793 customers. 1367 hours/7793 cus-tomers gives .175 hours per customer per year. Office No. 2 has 10379 customers. 2664/10379 gives .257 hours per customer. Obviously, office No. 2 is considerably more active than office No. 1.

The same calculation for office No. 3 produced 285 hours/ 2713 customers, or .105 hours/customer.

Office No. 4 turned out to be our busiest with 1776/4678, or .380 hours/customer.

Let's set our busiest office at unity activity. Then, office No. 1 is .175/.380, or 46 percent as busy per customer as office No. 4. Office No.2, .257/.380, or 68 percent. Office No. 3, .105/ .380, or 28 percent.

Office No. 4	1.00
Office No. 2	.68
Office No. 1	.46
Office No. 3	.28

That's quite a spread. But it is right in line with what we know about activity in these offices.

We have established our office No. 4 customer as our standard of maximum activity. Each one of these customers stirs up more dust than one customer of any other office.

Now if we moved all of office No. 1's customers out and replaced them with these troublemakers from office No. 4, it would increase activity at office No. 1 to 1.00/.46, or 217 percent of the present activity.

How many No. 4 customers could No. 1 stand without increasing the activity in No. 1? 46 percent of the present number of customers in office No 1.

How many "standard" (No.4) customers can each office stand without increasing its activity level? We can find out by multiplying the number of customers in each office by that office's activity index. As follows:

	Customers	Activity Index	Customers, Adjusted
Office No. 4	4678	1.00	4678
Office No. 2	10379	.68	7058
Office No. 1	7793	.46	3585
Office No. 3	2713	.28	760

Now, we can compare offices on the basis of customers, because we're using "standard" customers in each case. Let's see what the clerk and serviceman load is in each of our four offices, using " standard" customers as our base.

	Standard Customers	No. of Clerks	Customers per Clerk	No. of SVM	Customers per SVM
Office No. 4	4678	3	1559	2	2339
Office No. 2	7058	4	1765	4	1765
Office No. 1	3585	4.5	797	3	1195
Office No. 3	760	1	760	1	760

We see that the clerks in office No. 2 are carrying the heaviest load. The servicemen in office No. 4 are carrying the heaviest office-related load. Setting these two at unity, we can develop a percentage rating for each office which can be used as an accurate comparison of the relative "busyness" of these people.

	Clerk Load Rating			SVM Load Rating
Office No. 2	100	Office No. 4		100
Office No. 4	88	Office No. 2		75
Office No. 1	45	Office No. 1		51
Office No. 3	43	Office No. 3		32

The clerk in office No. 3 is faced with only 43 percent as much work load as the clerks in office No. 2. The serviceman in office No. 3 has only 32 percent as much work load which is directly related to office work as the serviceman in office No.4. He may, however, be doing more in the way of preventive maintenance, construction units, meter reading and lamp replacement than they are. Let's look at the E/R rating of the servicemen in each office compared to their load rating of purely office-related work.

	SVM Load Rating	E/R	Clerk Load Rating
Office No. 4	100	80	88
Office No. 2	75	56	100
Office No. 1	51	59	45
Office No. 3	32	65	43

What does this tell us?

The servicemen in office No. 4 are well loaded, but they can still take on a little more. The clerks are harder pressed than the servicemen.

The clerks in office No. 2 need some help. If not, they should be sent around to the other offices to train those clerks. We're overstaffed on servicemen or they are not being fully utilized. They're certainly not as busy as the clerks.

Office No. 1 has 4.5 clerks. Somebody has done a number on management. Something is badly out of whack. How can we support more than three clerks? We can't. We have one too many servicemen. Either that, or we need to dig up some work for them that they aren't doing now.

Office No. 3 has only one clerk and one serviceman. You can't reduce that. The serviceman is obviously being utilized a lot better than those in office No. 2. He could be used even better. And that clerk should enter her knitting in the state fair.

In using this data, we have to factor in lots of other information. Here we have a clerk nearing retirement. She does her best. She has to work late on her own time just to keep up and she never did quite grasp the new computer accounting methods. Compared to those sharp young teens and twenties gals in office No. 2, she's about 35 percent effective. So, we accept that and rate that office's performance accordingly. But when she retires, we shouldn't replace her with another .35. Maybe we shouldn't replace her at all.

Whatever we do, E/R will provide us with the basic standards on which we can make effective decisions in which we have complete confidence.

IMPLEMENT INDICATED IMPROVEMENTS

We've measured and we've evaluated. Again, we could stop right here. And here is where you're most likely to stop. Because, from here on, you must obtain agreement from higher levels of management to make any drastic changes indicated.

Of course, if the E/R rating system produces no definable, significant changes and improvements, enthusiasm will quickly wither and die. Timely, accurate reporting will become impossible. Morale will suffer and negative results will be forthcoming. It is imperative that the employees involved are continually convinced that the system is in active use for their benefit.

I can best illustrate by describing some actual examples of how the system has been applied in my company to effect improvements which were apparent to all employees involved.

From the E/R figures it became apparent that a three-man crew located in an outlying area comprising two local offices was being utilized ineffectively. Upon retirement of one of the two local managers we moved the crew to district headquarters where it received daily supervision as just three individuals, part of a ten-man work force, and we eliminated one serviceman and set up one manager to supervise both offices.

The servicemen remaining were now comfortably loaded with little time to sit around and complain about how overworked they were. The supervision was reduced to fit the company pattern for the number of customers involved. What little construction work was required was then accomplished with an occasional visit from a properly-sized work force and equipment from the district headquarters.

The customers were better served, the employees happier, and we saved about the equivalent of three employees on manpower.

Without the *facts* E/R gave us, this could never have been accomplished.

Another example involved a local office with two service-men. It is located near an army camp and is plagued with a continuous turnover in population. The number of customers, our only previous criterion of work load, did not indicate the need for any additional help. But E/R kept pointing to a work overload on both the servicemen and the office clerks. By stint of constantly holding these figures under management's nose we were able to get an additional clerk approved and we relieved the servicemen of the meter reading duties they carried by having the E/R underloaded district meter readers give that office several days of work a month.

How long this intolerable situation would have gone on without E/R facts on which to base our case, is anyone's guess.

Another major improvement resulted when E/R pointed up the fact that whereas the servicemen in most of our district headquarter local offices were fairly well loaded, most of our servicemen in outlying local offices were on a free ride. At the same time, most of our district meter readers had all they could say grace over. Solution? Let the local servicemen read as many of the meters in their own areas as was practical and redirect the activities of the district meter readers. This pointed to the ultimate elimination of almost half of our district meter reading jobs. Those men affected were transferred to the work force where they could aspire to higher wages than they could as meter readers.

In another case, E/R pointed to inadequate foreman supervision on a particular work group. Solution? We promoted a man to Foreman "B" to assist the single Foreman "A" supervise the eleven-man force. Morale improved markedly and efficiency followed.

In yet another case, a lone serviceman in a local office kept up a drumbeat about "running with his tongue hanging out." His tongue was perpetually in motion. I'm not surprised it lolled out on occasion for rest. E/R did not bear out that he should be all that busy. I sent a man to ride with him and trace his pattern for the day on an area map. It was a whore's dream. Looked like an atom trying to escape from a molecule. Traveled to one remote spot three times that day. A little supervision from his manager and he was able to put his tongue back in proper position.

The "A" Foreman of one construction group was condemned by his E/R ratings and after some study I insisted he be fired. His superintendent had a talk with him, he was "born-again," did a complete turnaround and E/R jumped like it was goosed.

One consistently high E/R performer—a serviceman—went into a slow dive. I suggested his manager check on his health. A "backache" turned out to be terminal cancer.

Does E/R incite undersirable, friction-promoting, safety-downgrading competition between individuals and groups? Remember, no individual ratings are published. As for groups, I have yet to find any evidence that one district's meter readers or servicemen feel impelled by poor E/R ratings to say, "Come on, boys, let's get with it. We've got to catch up with the other districts."

But they don't demand reinforcements.

There is some good, healthy competition between district superintendents who are in charge of the construction forces. It does not cause them to try to get their workers to hurry, but it does cause them to avoid bad work assignment procedures. They can't stand to be on the bottom for long. And you'll notice how closely the construction forces stay in E/R ratings compared to servicemen and meter readers.

Why does a particular serviceman or meter reader constantly score high on E/R? Send a man to ride with him for a day and report. Why does another serviceman or meter reader consistently score low? Send a man to ride with him and report.

Can you believe that one meter reader, with my man at his elbow, had his watch repaired on company time? And waited until it was finished? Another got a haircut. Why not? His supervisor hadn't laid out anything better for him to spend his time on.

The information provided by my "outrider" convinced me that we were sending as many as four men quite some distance to remote areas to perform tasks all of which could have been performed by one man. We were wearing out men and our trucks, meeting ourselves coming back, just for the lack of a little enlightened supervision.

No complaint against our workers. They were just doing what we had stupidly told them to do. But the hide is coming off those supervisors. They're supposed to know better. And they do. But they'd rather slide down a razor blade into a vat of alcohol than face their workers at 8:00 A.M. with a packet of tickets and say "Here's your day's work. Do it in the order that I've indicated. Now get going. I'll see you at 4:50 P.M."

E/R has spoken. Limit each man's day's work to the smallest area and have him perform all tasks in that area within his capability. Don't divert him during that day. That means that orders coming in today cannot be worked any sooner than tomorrow, unless they are genuine emergencies. It's as simple as that.

But you know it isn't that simple. Hear that noise? Foot-dragging. Excuses. Plain bellyaching.

It takes continued pressure, both up and down, to get old habits broken, old patterns changed. E/R is the lever that helps you pry supervisors and workers off their asses.

MONITOR THE EFFECTS

We've determined that a measure of efficiency is desirable and possible. We have developed a simple, accurate system. We've applied that system to warm bodies. We've evaluated the results and put the indicated changes into effect.

Once again, and for the last time, we can stop here.

This time, however, we will have accomplished a great deal and will not have generated any negative effects.

But if E/R is good for the workers and good for the supervisors, why shouldn't it be allowed to perform its magic on us, too? E/R has even measured its own success with a cold-blooded eye. Why not let it give us, its originators, the onceover? Let it measure our effectiveness in implementing the changes indicated by the statistics produced by E/R?

The truth is that we will not be at all loathe to do this. Our problem will arise in our impatience to prove our success via E/R too soon. We'll be tempted to make comparisons on one-month results comparing the month before we instituted our masterful personnel change or work procedure change with the first month the change is in effect.

No fair! You can't do that in making other comparisons. Too many variables. You can't do it now that you want a quick answer.

A twelve-month comparison is the only completely accurate comparison. Three months' accumulation is the very minimum time to produce a valid trend.

Start monitoring the change at three months and continue until you are into the change a full twelve months. Only then can you make an objective comparison. Most of the time you'll get the answer you want. But when you don't, admit you were wrong and take appropriate action.

What is more likely to happen is that if, after three months, you don't see the desired improvement reflected in the E/R rating, you will check and find out that the change you directed never was accomplished. You've been getting lots of talk and no action. And E/R has tattled on the bad little boys!

When E/R proves that a change you have instituted has produced desired improvement in performance, this information should be energetically utilized to effect other, similar changes.

The success which you have in using this system should not tempt you to try to use it on classifications of workers who do not spend the bulk of their time on repetitive tasks. It should not be applied to substation construction or maintenance. Other methods must be used for that.

Finally, the E/R rating system, and your managerial actions taken in accordance with the indicators supplied by the E/R statistics, should be evaluated in the light of cost per customer from your company's operating report. This should be compared against other divisions and it should be charted so that your performance today can be compared against past performance over the years.

Our excellent performance has shown steady improvement as measured by cost per customer and this improvement has paralleled the improvement measured by the E/R system.

I wish you equal success!

CONCLUSION

This book was not written as a service to mankind. It was written to make me a few bucks. Its service to mankind will be measured by the number of bucks it makes me.

This book will make you and your company lots more bucks than it cost. You're happy. I'm happy.

My division serves 90,000 customers. It is divided into five district organizations. It covers an area of 8,000 square miles.

You may question the value to your operation of my methods which were developed for an operation the size of mine. I answer that all utility operations, no matter how vast, are made up of segments the size of my operation. But those segments may not resemble my operation in geography, organization or personnel. So you can't apply my methods to such an organization in the same way I have applied my methods to my organization. They must be tailored to your situation.

But the basic concepts are valid for any size operation. Trees is trees. People is people. Efficiency is efficiency. Lightning is lightning.

One more thing. In discussing each area of operation I have concentrated on only what I have done *differently* from what I have observed others doing. I have not bored you with a long-winded, detailed description of each phase of operation in its entirety. You already know all that.

You know a lot more about a lot of utility operations than I do. I just hope you know even more now that you've read my book.